MLOps Lifecycle Toolkit

A Software Engineering Roadmap for Designing, Deploying, and Scaling Stochastic Systems

Dayne Sorvisto

Apress®

MLOps Lifecycle Toolkit: A Software Engineering Roadmap for Designing, Deploying, and Scaling Stochastic Systems

Dayne Sorvisto
Calgary, AB, Canada

ISBN-13 (pbk): 978-1-4842-9641-7   ISBN-13 (electronic): 978-1-4842-9642-4
https://doi.org/10.1007/978-1-4842-9642-4

Managing Director, Apress Media LLC: Welmoed Spahr
Acquisitions Editor: Celestin Suresh John
Development Editor: Laura Berendson
Editorial Assistant: Mark Powers

Cover designed by eStudio Calamar

Cover image by Jerry Yaar on Pixabay (www.pixabay.com)

Distributed to the book trade worldwide by Springer Science+Business Media New York, 1 New York Plaza, Suite 4600, New York, NY 10004-1562, USA. Phone 1-800-SPRINGER, fax (201) 348-4505, e-mail orders-ny@springer-sbm.com, or visit www.springeronline.com. Apress Media, LLC is a California LLC and the sole member (owner) is Springer Science + Business Media Finance Inc (SSBM Finance Inc). SSBM Finance Inc is a **Delaware** corporation.

For information on translations, please e-mail booktranslations@springernature.com; for reprint, paperback, or audio rights, please e-mail bookpermissions@springernature.com.

Apress titles may be purchased in bulk for academic, corporate, or promotional use. eBook versions and licenses are also available for most titles. For more information, reference our Print and eBook Bulk Sales web page at http://www.apress.com/bulk-sales.

Any source code or other supplementary material referenced by the author in this book is available to readers on GitHub (github.com/apress). For more detailed information, please visit http://www.apress.com/source-code.

Printed on acid-free paper

I dedicate this book to my mom, my wife, and Lucy.

Table of Contents

About the Author

 **Dayne Sorvisto** has a Master of Science degree in Mathematics and Statistics and became an expert in MLOps. He started his career in data science before becoming a software engineer. He has worked for tech start-ups and has consulted for Fortune 500 companies in diverse industries including energy and finance. Dayne has previously won awards for his research including the Industry Track Best Paper Award. He has also written about security in MLOps systems for Dell EMC's Proven Professional Knowledge Sharing platform and has contributed to many of the open source projects he uses regularly.

About the Technical Reviewer

 **Ashutosh Parida** is an accomplished leader in artificial intelligence and machine learning (AI/ML) and currently serving as Assistant Vice President, heading AI/ML product development at DeHaat, a leading AgriTech start-up in India. With over a decade of experience in data science, his expertise spans various domains, including vision, NLU, recommendation engines, and forecasting.

With a bachelor's degree in Computer Science and Engineering from IIIT Hyderabad and a career spanning 17 years at global technology leaders like Oracle, Samsung, Akamai, and Qualcomm, Ashutosh has been a site lead for multiple projects and has launched products serving millions of users. He also has open source contributions to his credit.

Stay connected with Ashutosh on LinkedIn to stay updated on his pioneering work and gain valuable industry insights: linkedin.com/in/ashutoshparida.

Acknowledgments

I want to thank my wife, Kaye, for providing me with boundless inspiration throughout the writing of this book. I also want to thank the entire Apress team from the editors that turned my vision for this book into a reality to the technical reviewer, project coordinators, and everyone else who gave feedback and contributed to this book. It would not be possible without you.

Introduction

This book was written in a time of great change in data science. From generative AI, regulatory risks to data deluge and technological change across industries, it is not enough to just be data-savvy. You increasingly need to understand how to make technical decisions, lead technical teams, and take end-to-end ownership of your models. *MLOps Lifecycle Toolkit* is your pragmatic roadmap for understanding the world of software engineering as a data scientist.

In this book I will introduce you to concepts, tools, processes, and labs to teach you MLOps in the language of data science. Having had the unique experience of working in both data science and software engineering, I wrote the book to address the growing gap I've observed first-hand between software engineers and data scientists. While most data scientists have to write code, deploy models, and design pipelines, these tasks are often seen as a chore and not built to scale. The result is increased technical debt and failed projects that threaten the accuracy, validity, consistency, and integrity of your models.

In this book you will build your own MLOps toolkit that you can use in your own projects, develop intuition, and understand MLOps at a conceptual level. The software toolkit is developed throughout the book with each chapter adding tools that map to different phases of the MLOps lifecycle from model training, model inference and deployment to data ethics. With plenty of industry examples along the way from finance to energy and healthcare, this book will help you make data-driven technical decisions, take control of your own model artifacts, and accelerate your technical roadmap.

Source Code

All source code used in this book can be downloaded from `github.com/apress/mlops-lifecycle-toolkit`.

CHAPTER 1

Introducing MLOps

As data scientists we enjoy getting to see the impact of our models in the real world, but if we can't get that model into production, then the data value chain ends there and the rewards that come with having high-impact research deployed to production will not be achieved. The model will effectively be dead in the model graveyard, the place where data science models go to die.

So how do we keep our models out of this model graveyard and achieve greater impact? Can we move beyond simply measuring key performance indicators (KPIs) to moving them so that our models become the driver of innovation in our organization? It's the hypothesis of this book that the answer is yes but involves learning to become better data-driven, technical decision makers. In this chapter, I will define MLOps, but first we need to understand the reasons we need a new discipline within data science at all and how it can help you as a data scientist own the entire lifecycle from model training to model deployment.

What Is MLOps?

Imagine you are the director of data science at a large healthcare company. You have a team of five people including a junior data analyst, a senior software (data) engineer, an expert statistician, and two experienced data scientists. You have millions of data sets, billions of data points from thousands of clinical trials, and your small team has spent the last several sprints developing a model that can change real people's lives.

© Dayne Sorvisto 2023
D. Sorvisto, *MLOps Lifecycle Toolkit*, https://doi.org/10.1007/978-1-4842-9642-4_1

You accurately predict the likelihood that certain combinations of risk factors will lead to negative patient outcomes, predict the posttreatment complication rate, and you use an inflated Poisson regression model to predict the number of hospital visits based on several data sources. Your models are sure to have an impact, and there's even discussion about bringing your research into a convolutional neural network used to aid doctors in diagnosing conditions. The model you created is finally at the cutting edge of preventative medicine. You couldn't be more excited, but there's a problem.

After several sprints of research, exploratory data analysis (EDA), data cleaning, feature engineering, and model selection, you have stakeholders asking some tough questions like, When is your model going to be in production? All of your code is in a Jupyter notebook, your cleaning scripts scattered in various folders on your laptop, and you've done so many exploratory analyses you're starting to have trouble organizing them.

Then, the chief health officer asks if you can scale the model to include data points from Canada and add 100 more patient features to expand into new services all while continuing your ad hoc analysis for the clinical trial. By the way, can you also ensure you've removed all PII from your thousands of features and ensure your model is compliant with HIPAA (Health Insurance Portability and Accountability Act)? At this point, you may be feeling overwhelmed.

As data scientists we care about the impact our models have on business, but when creating models in the real world, the process of getting your model into production so it's having an impact and creating value is a hard problem. There are regulatory constraints; industry-specific constraints on model interpretability, fairness, data science ethics; hard technical constraints in terms of hardware and infrastructure; and scalability, efficiency, and performance constraints when we need to scale our models to meet demand and growing volumes of data (in fact, each order of magnitude increase in the data volume leads to new architectures entirely).

MLOps can help you as a data scientist take control of the entire machine learning lifecycle end to end. This book is intended to be a rigorous approach to the emerging field of ML engineering, designed for the domain expert or experienced statistician who wants to become a more end-to-end data scientist and better technical decision maker.

The plan then is to use the language of data science, examples from industries and teach you the tools to build ML infrastructure; deploy models; set up feature groups, training pipelines, and data drift detection systems to accelerate your own projects; comply with regulatory standards; and reduce technical debt. Okay, so with this goal in mind, let's teach you the MLOps lifecycle toolkit, but first let us take the first step in a million-mile journey and define exactly what we mean by MLOps.

Defining MLOps

We need to define MLOps, but this is a problem because at the present time, MLOps is an emerging field. A definition you'll often hear is that MLOps is the intersection of people, processes, and technology, but this definition lacks specificity. What kind of people? Domain experts? Are the processes in this definition referring to DevOps processes or something else, and where in this definition is data science? Machine learning is but one tool in the data scientist's toolkit, so in some sense, MLOps is a bit of a misnomer as the techniques for deploying and building systems extend beyond machine learning (you might think of it as "Data Science Ops").

In fact, we are also not talking about one specific industry where all of the MLOps techniques exist. There is no one industry where all of the data science is invented first, but in fact each industry may solve data science problems differently and have its own unique challenges for building and deploying models. We revised the definition and came up with the following definition that broadly applies to multiple industries and also takes into account the business environment.

MLOps definition: MLOps (also written as ML Ops) is the intersection of industry domain experts, DevOps processes, and technology for building, deploying, and maintaining reliable, accurate, and efficient data science systems within a business environment. Figure 1-1 illustrates this.

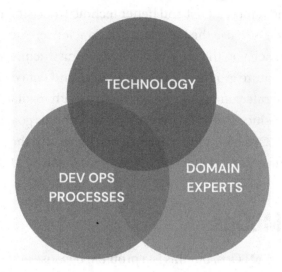

Figure 1-1. *Stakeholders will never trust a black box*

In machine learning, we solve optimization problems. We have data, models, and code. Moreover, the output of models can be non-deterministic with stochastic algorithms that are difficult to understand or communicate due to their mathematical nature and can lead stakeholders to viewing the system as an opaque "black box." This view of a machine learning system as a black box is a real barrier to trusting the system and ultimately accepting its output. If stakeholders don't trust your model, they won't use your model. MLOps can also help you track key metrics and create ***model transparency***.

This is the reason machine learning in the real world is a hard problem. Oftentimes the models themselves are a solved problem. For example, in the transportation sector, we know predicting the lifetime of a component in a fleet of trucks is a regression problem. For other problems, we may have many different types of approaches and even cutting-edge research that has not been battle-tested in the real world, and as a machine learning engineer, you may be the first to operationalize an algorithm that is effectively a black box, which not only is not interpretable but may not be reliable. We need to be transparent about our model output and what our model is doing, and the first step is to begin measuring quality and defining what success means in your project.

If stakeholders don't trust your model, they won't use your model.

So how do we even begin to measure the quality of an MLOps solution when there is so much variability in what an MLOps solution looks like? The answer to this conundrum is the MLOps maturity model, which applies across industries.

MLOps Maturity Model

What does it mean to have "good" MLOps practices, and what is the origin of this term? You probably have several questions around MLOps and might be wondering what the differences are between MLOps and software development, so let us first discuss, in the ideal case, what MLOps looks like by presenting the maturity model in diagram form (Figure 1-2).

MLOps Maturity Model

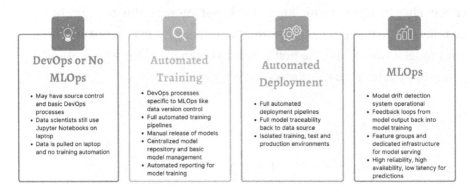

Figure 1-2. *MLOps maturity model*

Notice that the key differentiator from DevOps to stage 2 MLOps is the existence of an automated training pipeline with automated reporting requiring data management and infrastructure for reporting. The key differentiator from stage 2 to stage 3 is the existence of automated release pipelines (model management), and the differentiator between stage 3 and the final stage (full-blown MLOps) is we approach high reliability including specific subsystems to detect data and model drift and some way of creating feedback required to "move the needle" on key measurements we'll cover.

Brief History of MLOps

MLOps has its history in DevOps processes. DevOps, the merging of the words "Dev" and "Ops," changed the way software developers built, deployed, and maintained large-scale software systems. DevOps includes best practices for automating, continuous delivery, software testing and emphasizes developing software for the end user. However, if you notice the diagram of the MLOps maturity model, DevOps is not the same as MLOps, and it is not sufficient to apply DevOps principles to data science without some adaptation.

MLOps in contrast is about continuous delivery of machine learning systems but extends this idea to continuous training and reproducibility (an important part of the scientific process).

You may be thinking, well, software development is software development, which naturally encompasses parts of the software development lifecycle including infrastructure as code (creating reusable environments), containerization, CI/CD pipelines, and version and model control systems. However, this stuff does not automatically give you model management and data management best practices. Hence, MLOps takes DevOps a step further and can be thought of as a multi-dimensional version of DevOps that fuses together best practices from software engineering, ModelOps, and DataOps to move key metrics such as interpretability, accuracy, reliability, and correlation with key performance indicators specific to your industry.

Defining the Relationship Between Data Science and Engineering

Okay, we have defined MLOps, but it's important we have a clear idea on what we mean by data science and software engineering and the differences between them.

For the purpose of this book, we use data science as an umbrella term for a multidisciplinary approach to extracting knowledge from data that uses the scientific method. Data science uses machine learning as a tool. This is why we talk about stochastic systems instead of just machine learning systems since MLOps practices can apply to more than just machine learning, for example, operationalizing something more abstract like a causal model or Bayesian network or custom statistical analysis. In the next section, we will look at some general patterns for data science projects.

What Are the Types of Data Science Projects?

It is vital to understand the types of data science projects you might encounter in the real world before we dive into the *MLOps lifecycle*, which will be the focus of the rest of the book. We will look at supervised machine learning, semi-supervised machine learning, reinforcement learning, probabilistic programming paradigms, and statistical analysis.

Supervised Machine Learning

Supervised machine learning is a machine learning problem that requires labeled input data. Examples of supervised learning include classification and regression. The key is labeled data. For example, in NLP (natural language processing), problems can be supervised. You might be building a classification model using a transformer architecture to recommend products for new customers based on free-form data. Although you might know how to build a transformer model in PyTorch or TensorFlow, the challenge comes from labeled data itself. How do you create labels? Furthermore, how do you ensure these labels are consistent? This is a kind of chicken and egg problem that machine learning teams often face, and although there are solutions like Mechanical Turk, with privacy and data regulations like GDPR, it may be impossible to share sensitive data, and so data needs to be labeled internally, creating a kind of bottleneck for some teams.

Semi-supervised Machine Learning

In semi-supervised problems, we have a rule for generating labels but don't explicitly have labeled data. The difference between semi-supervised algorithms is the percentage of training data that is unlabeled. Unlike supervised learning where consistency of labels may be an issue, with semi-supervised the data may consist of 80% unlabeled data and a small percentage, say 20%, of labeled data. In some cases like fraud detection in banking, this is a very important class of machine learning problems since not all cases of fraud are even known and identifying new cases of fraud is a semi-supervised problem. Graph-based semi-supervised algorithms have been gaining a lot of traction lately.

Reinforcement Learning

Reinforcement learning is a type of machine learning where the goal is to choose an action that maximizes a future reward. There are several MLOps frameworks for deploying this type of machine learning system such as Ray, but some of the challenge is around building the environment itself, which may consist of thousands, millions, or billions of states depending on the complexity of the problem being modeled. We can also consider various trade-offs between exploration and exploitation.

Probabilistic Programming

Frameworks like PyMC3 allow data scientists to create Bayesian models. Unfortunately, these models tend not to be very scalable, and so this type of programming for now is most often seen during hyper-parameter tuning using frameworks like Hyperopt where we need to search over a large but finite number of hyper-parameters but want to perform the sweep in a more efficient way than brute force or grid search.

Ad Hoc Statistical Analysis

You may have been asked to perform an "EDA" or exploratory data analysis before, often the first step after finding suitable data sources when you're trying to discover a nugget of insight you can act upon. From an MLOps perspective, there are important differences between this kind of "ad hoc" analysis and other kinds of projects since operationalizing an ad hoc analysis has unique challenges.

One challenge is that data scientists work in silos of other statisticians creating ad hoc analysis. Ad hoc analysis is kind of an all-in category for data science code that is one-off or is not meant to be a part of an app or product since the goal may be to discover something new. Ad hoc analysis can range from complex programming tasks such as attribution modeling to specific statistical analysis like logistic regression, survival analysis, or some other one-off prediction.

Another noteworthy difference between ad hoc analysis and other types of data science projects is ad hoc analysis is likely entirely coded in a Jupyter notebook either in an IDE for data scientists such as Anaconda in the case the data scientist is running the notebook locally or Databricks notebook in the cloud environment collaborating with other data scientists.

An ad hoc analysis may be a means to an end, written by a lone data scientist to get an immediate result such as estimating a population parameter in a large data set to communicate to stakeholders.

Examples of ad hoc analysis might include the following:

- Computing correlation coefficient

- Estimating feature importance for a response variable in an observational data set

- Performing a causal analysis on some time series data to determine interdependencies among time series components

- Visualizing a pairwise correlation plot for variables in your data set to understand dependence structure

The Two Worlds: Mindset Shift from Data Science to Engineering

It is no secret that data science is a collaborative sport. The idea of a lone data scientist, some kind of mythical persona that is able to work in isolation to deliver some groundbreaking insight that saves the company from a disaster using only sklearn, probably doesn't happen all that often in a real business environment. Communication is king in data science; the ability to present analysis and explain what your models are doing to an executive or a developer requires to shift mindsets and understand your problem from at least two different perspectives: the technical perspective and the business perspective. The business perspective is talked about quite a bit, how to communicate results to stakeholders, but what about the other half of this? Communicating with other technical but non–data science stakeholders like DevOps and IT?

The topic of cross-team communication in data science crops up when requesting resources, infrastructure, or more generally in any meetings with non–data scientists such as DevOps, IT, data engineering, or other engineering-focused roles as a data scientist.

Leo Breiman, the creator of random forests and bootstrap aggregation (bagging), wrote an essay entitled "Statistical Modeling: The Two Cultures." Although Breiman may not have been talking about type A and type B data scientists specifically, we should be aware that in a multidisciplinary field like data science, there's more than one way to solve a problem and sometimes one approach, although valid, is not a good culture fit for every technical team and needs to be reframed.

What Is a Type A Data Scientist?

Typically a type A data scientist is one with an advanced degree in mathematics, statistics, or business. They tend to be focused on the business problem. They may be domain experts in their field or statisticians (both frequentist or Bayesian), but they might also come from an applied math or business background and be non-engineering.

These teams may work in silos because there is something I'm going to define as the *great communication gap* between type A and type B data scientists. The "B" in type B stands for *building* (not really, but this is how you can remember the distinction).

As data science matures, the distinction may disappear, but more than likely data scientists will split into more specialized roles such as data analyst, machine learning engineer, data engineer, and statistician, and it will become even more important to understand this distinction, which we present in Table 1-1.

Table 1-1. *Comparing Type A and Type B Data Scientists*

Type A Data Scientist	Type B Data Scientist
Focuses on understanding the process that generated the data	Focuses on building and deploying models
Focuses on measuring and defining the problem	Focuses on building infrastructure and optimizing models
Values statistical validity, accuracy, and domain expertise	Values system performance, efficiency, and engineering expertise

Types of Data Science Roles

Over the past decade, data science roles have become more specialized, and we often see roles such as data analyst, data engineer, machine learning engineer, subject matter expert, and statistician doing data science work to address challenges. Here are the types of data science roles:

- *Business analysts*: Problems change with the market and model output (called data drift or model drift). The correlation of model output with key business KPIs needs to be measured, and this may require business analysts who understand what the business problem means.

- *Big data devs*: Data volume can have properties such as volume, veracity, and velocity that transactional systems are not designed to address and require specialized skills.

- *DevOps*: Data needs to be managed as schemas change over time (called schema drift) creating endless deployment cycles.

- *Non-traditional software engineers*: Data scientists are often formally trained in statistics or business and not software engineering.

Even within statistics, there is division between Bayesians and frequentists. In data science there are also some natural clusters of skills, and often practitioners have a dominant skill such as software engineering or statistics.

Okay, so there's a rift even within statistics itself, but what about across industries? Is there one unified "data scientist"?

For example, geospatial statistics is its own beast with spatial dependence of the data unlike most data science workflows, and in product companies, R&D data scientists are highly sought after as not all model problems are solved and they require iterating on research and developing reasoning about data from axioms. For example, a retail company may be interested in releasing a new product that has never been seen on the market and would like to forecast demand for the product. Given the lack of data, novel techniques are required to solve the "cold start" problem. Recommender systems, which use a collaborative filtering approach to solve this problem, are an example, but oftentimes out-of-the-box or standard algorithms fall short. For example, slope-one, a naive collaborative filtering algorithm, has many disadvantages.

Hackerlytics: Thinking Like an Engineer for Data Scientists

The ability to build, organize, and debug code is an invaluable skill even if you identify as a type "A" data scientist. Automation is a key ingredient in this mindset, and we will get there (we cover specific MLOps tools like PySpark, MLflow, and give an introduction to programming in Python and Julia in the coming chapters), but right now we want to focus on concepts. If you understand the concept of technical debt, which is particularly relevant in data science, and the need to future-proof your code, then you will appreciate the MLOps tools and understand how to use them without getting bogged down in technical details. In order to illustrate the concept of technical debt, let's take a look at a specific example that you might have encountered when building a machine learning pipeline with real data.

Anti-pattern: The Brittle Training Pipeline

Suppose you work for a financial institution where you're asked by your data science lead to perform some data engineering task like writing a query that pulls in the last 24 months of historical customer data from an analytical cloud data warehouse (the database doesn't matter for this example; it could be anything like a SQL Pool or Snowflake). The data will be used for modeling consumer transactions and identifying fraudulent transactions, which are only 0.1% of the data.

You need to repeat this process of extracting customer transaction data and refreshing the table weekly from production so you have the latest to build important features for each customer like number of recent chargebacks and refunds. You are now faced with a technical choice: do you build a single table, or do you build multiple tables, one for each week that may make it easier for you to remember?

You decide to opt for this latter option and build one table per week and adjust the name of the table, for example, calling it something such as historical_customer_transactions_20230101 for transaction dates ending on January 1, 2023, and the next week historical_customer_transactions_20230108 for transactions ending on January 8, 2023. Unfortunately, this is a very brittle solution and may not have been a good technical decision.

What is brittleness? Brittleness is a concept in software engineering that is hard to grasp without experiencing its *future consequences.* In this scenario, our solution is brittle because a single change can break our pipelines or cause undue load on IT teams. For example, within six months you will have around 26 tables to manage; each table schema will need to be source controlled, leading to 26 changes each time a new feature is added. This could quickly become a nightmare, and building training pipelines will be challenging since you'll need to store an array of dates and think about how to update this array each time a new date is added. So how do we fix this?

If we pick the first option, a single table, can we make this work and eliminate the array of dates from our training pipeline and reduce the effort it takes to manage all of these tables? Yes, easily in this case we can add metadata to our table, something like a snapshot date, and give our table a name that isn't tethered to a specific datetime, something like historical_customer_transaction (whether your table name is plural or singular is also a technical decision you should establish early in your project). Understanding, evaluating, and making technical decisions like this comes with experience, but you can learn to become a better technical decision maker by applying our first MLOps tool: *future-proofing your code.*

Future-Proofing Data Science Code

As we discussed, a better way to store historical transaction data is to add an additional column to the table rather than in the table name itself (which ends up increasing the number of tables we have to manage and thus technical debt, operational risk, and weird code necessary to deal with the decision such as handling an unnecessary dynamic array of dates in a training pipeline).

From a DevOps perspective, this is fantastic news because you will reduce the IT load from schema change and data change down to a simple insert statement.

As you develop an engineering sense should be asking two questions before any technical decision:

- Am I being consistent? (Example: Have I used this naming convention before?)

- If I make this technical decision, what is the future impact on models, code, people, and processes?

Going back to our original example, by establishing a consistent naming convention for tables and thinking about how our naming convention might impact IT that may have to deploy 26 scripts to refresh a table, if we choose a poor naming convention such as table sprawl, code spiral, or repo sprawl, we'll start to see cause and effect relationships and opportunities to improve our project and own workload as well. This leads us to the concept of ***technical debt***.

What Is Technical Debt?

"Machine learning is the high interest credit card of technical debt."[1] Simply put, technical debt occurs when we write code that doesn't anticipate change. Each suboptimal technical decision you make now doesn't just disappear; it remains in your code base and will at some point, usually at the worst time (Murphy's Law), come back to bite you in the form of crashed pipelines, models that choke on production data, or failed projects.

Technical debt may occur for a variety of reasons such as prioritizing speed of delivery over all else or a lack of experience with basic software engineering principles such as in our brittle table example. To illustrate the concept of technical debt and why it behaves like real debt, let's consider another industry-specific scenario.

Imagine you are told by the CEO of a brick-and-mortar retail company that you need to build a model to forecast customer demand for a new product. The product is similar to one the company has released before, so there is data available, but the goal is to use the model to reduce costs of storing unnecessary product inventory. You know black box libraries won't be sufficient and you need to build a custom model and feature engineering library.

Your engineering sense is telling you that building a custom solution will require understanding various trade-offs. Should you build the perfect model and aim for 99% accuracy and take a hit on performance? Does the business need 99% accuracy, or will forecasting demand with 80% accuracy be sufficient to predict product inventory levels two weeks in advance?

---

[1] Machine Learning: The High Interest Credit Card of Technical Debt

Hidden Technical Trade-Offs in MLOps

In the previous example, we identified a ***performance-accuracy trade-off*** (Figure 1-3) that is one of many trade-offs you'll face in technical decision making when wearing an MLOps hat. MLOps is full of these hidden technical trade-offs that underlie each technical decision you make. By understanding the kinds of trade-offs, you can reduce technical debt instead of accumulating it. We've summarized some common trade-offs in data science:

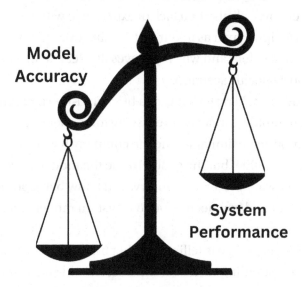

Figure 1-3. *Data science projects have many hidden technical trade-offs*

- Data volume vs. model accuracy (more data is better, but each 10× scale-up requires new infrastructure)

- Technical correctness vs. cognitive complexity (data science code has high cognitive complexity especially when handling every possible edge case, which can cause performance bottlenecks)

- Model accuracy vs. model complexity (do you really need
to use deep learning, or is a decision tree–based model
that is 90% accurate sufficient for the first iteration?)

How to Protect Projects from Change

Change is everywhere! Change is also inevitable and can be the seed
of innovation in data science, so it has its upsides and downsides. The
downsides are change can increase technical debt and, if not properly
managed, can cause failed data science projects.

So where does it come from? What are the main drivers of change in
data science? Change can come from stakeholders or market conditions,
problem definition, changes in how a KPI is measured, or schema changes.
You need to learn to protect your data science projects from change.

Software engineering principles are about learning to protect against
change and be future thinking and so can be applied to data science.
Writing clean code has little to do with being a gatekeeper or annoyance
but is a necessary part of building a reliable system that isn't going to
crash on Sunday's nightly data load and cause your team to be called in to
troubleshoot an emergency.

Maybe the most basic example of shielding from change in data
science is the concept of a view. A view is an abstraction, and as software
engineers we like abstractions since they allow us to build things and
depend on something that is stable and unchanging such as the name of a
view, even if what is being abstracted, the query itself, the schema, and the
data underneath, is constantly changing.

Managing views, source control, and understanding when to
apply abstractions are something that can come with practice, but
understanding the value of a good abstraction will take you a long way
in shielding your own code from change and understanding some of the
reasons technical decisions are made without becoming frustrated in the
process.

There are abstractions for data like views we can use to manage changes in data along with other tools like data versioning, but there are also abstractions for models and code like source control and model registries like MLflow. We'll talk about all of these MLOps tools for managing change in subsequent chapters, but keep in mind the concept of abstractions and how these tools help us protect our project from change.

Drivers of Change in Data Science Projects

We know the types of approaches needed to build an attribution model, but there is no one way to build one without historical data, and the types of approaches may involve more than something you're familiar with like semi-supervised learning; it may involve, instead, stochastic algorithms or a custom solution. For attribute modeling we could think about various techniques from Markov chains to logistic regression of Shap values.

From a coding perspective, for each type of approach, we are faced with a choice as a designer of a stochastic learning system on programming language, framework, and tech stack to use. These technologies exist in the real world and not in isolation, and in the context of a business environment, change is constantly happening.

These combinatorial and business changes, called change management, can cause disruptions and delays in project timelines, and for data science projects, the line is often a gap between what the business wants and the problem that needs well-defined requirements or at worst an impossible problem or one that would require heroic efforts to solve within the time and scope.

So model, code, well-defined requirements... What about the data? We mentioned the business is constantly changing, and this is reflected in the data and the code. It is often said that a company ships its own org chart, and the same is true for data projects where changes to business entities cause changes in business rules or agreement upon ways to measure

KPIs for data science projects, which leads to intense downstream change to feature stores, feature definitions, schema changes, and downstream pipelines for model training and inference.

Externalities or macro-economic conditions may also cause changes in customer assumptions and behavior that get reflected in the model, a problem often called concept drift. These changes need to be monitored and acted upon (e.g., can we retrain the model when we detect concept drift), and these actions need to be automated and maintained as packages of configuration, code, infrastructure as code, data, and models. Artifacts like data and models require versioning and source control systems, and these take knowledge of software engineering to set up.

Choosing a Programming Language for Data Science

Arguments over programming languages can be annoying, especially when this leads to a holy war instead of which programming language is the right tool for the job. For example do you want performance, type safety, or the ability to rapidly prototype with extensive community libraries?

This is not a programming book, and it's more important that you learn how to think conceptually, which we will cover in the next chapter. Python is a very good language to learn if you are starting out, but there are other languages like Julia and R and SQL that each have their uses. You should consider the technical requirements, skill set of your team before committing to a language. For example, Python has distinct advantages over R for building scalable code, but when it comes to speed, you might consider an implementation in Julia or C++. This also comes with a cost: the data science community for Python is prolific, and packages like Pandas and sklearn are widely supported. This isn't to say using a language like R or Julia is wrong, but you should make a decision based on available data.

More advanced data scientists interested in specializing in MLOps may learn C++, Julia, or JAX (for accelerating tensor operations) in addition to Python and strong SQL skills.

We'll cover programming basics including data structures and algorithms in the following chapter. It's worth noting that no one language is best and new languages are developed all the time. In the future, functional programming languages oriented around probabilistic programming concepts will likely play a bigger role in MLOps.

MapReduce and Big Data

Big data is a relative term. What was considered "big data" 20 years ago is different from what is considered "big data" today. The term became popular with the advent of MapReduce algorithms and Google's Bigtable technology, which allowed algorithms that could be easily parallelized to be run over extremely large gigabyte-scale data sets. Today, we have frameworks like Spark, and knowledge of MapReduce and Java or Scale isn't necessary since Spark has a Python API called PySpark and abstracts the concept of a mapper and reducer from the user. However, as data scientists we should understand when we're dealing with "big data" as there are specific challenges. Not all "big data" is high volume. In fact there are three Vs in big data you may need to handle.

Big Data a.k.a. "High Volume"

This is most commonly what is meant by "big data." Data that is over a gigabyte may be considered big data, but it's not uncommon to work with billions of rows or terabytes of data sourced from cold storage. High-volume data may pose operational challenges in data science since we need to think about how to access it and transferring large data sets can be a problem. For example, Excel has a famous 1 million row limit, and similarly with R, the amount of memory is restricted to 1 GB, so we need tools like Spark to read and process this kind of data.

High-Velocity Data

By "high-velocity" data sources, we usually mean streaming data. Streaming data is data that is unbounded and may come from an API or IoT edge device in the case of sensor data. It's not the size of the data that is an operational challenge but the speed at which data needs to be stored and processed. There are several technologies for processing high-velocity data including "real-time" or near-real-time ones (often called micro-batch architecture) like Apache Flink or Spark Streaming.

High-Veracity Data

If you are a statistician, you know the concept of variability. Data can have variability in how it's structured as well. When data is unstructured like text or semi-structured like JSON or XML in the case of scraping data from the Web, we refer to it as high veracity. Identifying data sources as semi-structured, structured, or unstructured is important because it dictates which tools we use and how we represent the data on a computer, for example, if dealing with structured data, it might be fine to use a data frame and Pandas, but if our data is text, we will need to build a pipeline to store and process this data and you may even need to consider NoSQL databases like MongoDB for this type of problem.

Types of Data Architectures

We might choose a data architecture to minimize change like a structured data warehouse, but when it comes to data science projects, the inherent inflexibility of a structured schema creates a type of impedance mismatch. This is why there are a number of data architectures such as the data lake,

23

data vault, or medallion architecture that may be better fit for data science. These architectures allow for anticipated changes in schema, definition, and business rules.

The Spiral MLOps Lifecycle

We will discuss the titular MLOps lifecycle in detail in Chapter 7, but we can broadly distinguish the different phases of a data science project into

1. Insight and data discovery

2. Data and feature engineering

3. Model training and evaluation

4. Model deployment and orchestration

The reason we call it a spiral lifecycle is because each of these stages may feedback into previous stages; however, as the technical architecture matures (according to the MLOps maturity model), the solution should converge to one where you are delivering continuous value to stakeholders. Figure 1-4 shows the spiral MLOps lifecycle.

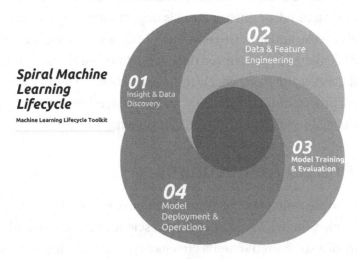

Figure 1-4. *The spiral MLOps lifecycle*

Data Discovery

Ideally, you would approach the problem like a statistician. You have a problem, design an experiment, choose your sampling method, and collect the exact data you need to ensure there are no biases in your data, repeating the process as many times as you can if need be. Unfortunately, business does not work like this. You will be forced to approach the problem backward; you might have a vague idea of what problem you want to solve and the data you need, but you may only have the resources or access to the data that the business has already collected.

How can MLOps help with this process? Well, the data is often stored as flat CSV files, which can total several gigabytes or more. A typical scenario is you have to figure out how to load this data and you quickly run into memory errors if using tools like Pandas. We'll show you how to leverage distributed computing tools like Databricks to get around these issues without too much headache and if possible without even rewriting your code.

Data Discovery and Insight Generation

This phase is all about exploring your data sets, forming and testing hypotheses, and developing intuition for your data. There are often several challenges at this stage. If you are using a tool like seaborn or matplotlib for generating visualizations, you might face the challenge of how to deploy your work or share it with other data scientists or stakeholders.

If you're working on an NLP problem, you might have lots of different experiments and want to quickly test them. How do you organize all of your experiments, generate metrics and parameters, and compare them at different points in time? Understanding standard tools like MLflow and how to set up an experimentation framework can help.

Let us suppose as a data scientist you are an expert at understanding and identifying biases in your data. You work for a financial institution and are tasked with creating a customer churn model. You know net promoter score is a key metric, but you notice survey responses for your customers are missing on two key dates.

You decide to write a script in Pandas to filter out these two key dates. Suddenly, the next week your data has doubled in size, and your script no longer scales. You manually have to change the data in a spreadsheet and upload it to a secure file server. Now, you spend most of your time on this manual data cleaning step, validating key dates and updating them in a spreadsheet. You wish you knew how to automate some of these steps with a pipeline.

Few people enjoy doing ETL or ELT; however, understanding the basics of building pipelines and when to apply automation at scale especially during the data cleaning process can save you time and effort.

Data and Feature Engineering

Feature selection may be applied to both supervised and unsupervised machine learning problems. In the case that labeled data exists, we might use some measure like correlation to measure the degree of independence between our features and a response or target variable. It may make sense to remove features that are not correlated with the target variable (a kind of low-pass filter), but in other cases, we may use a model itself like Lasso or Ridge regression or random forest to decide which features are important.

Model Training

How do you choose the best model for your problem? A training pipeline in the wild can take hours or even sometimes days to finish especially if the pipeline consists of multiple steps.

As an MLOps engineer, you should be familiar with frameworks and methods for speeding up model training. We'll look at Hyperopt, a framework for using Bayesian hyperparameter search, and Horovod for distributed model training that takes advantage of parallelism. By speeding up model training time, by using distributed computing or GPU, we can immediately add value to a project and have more time spent doing data science.

Model Evaluation

Model selection is the process of choosing the "best" model for a business problem. Best may not necessarily be as simple as the model with the best training accuracy in case the data overfits and does not generalize to new samples (see the bias-variance trade-off). It may be more nuanced than this, as there might be regulatory constraints such as in the healthcare and banking industries where model interpretability and fairness are a concern. In this case, we must make a technical decision, balancing the attributes of the model we desire like precision, recall, and accuracy over how interpretable or fair the model is and what kind of data sources we are legally allowed to use to train the model. MLOps can help with this step of the machine learning lifecycle by automating the process of model selection and hyper-parameter tuning.

Deployment and Ops

You've trained your model and even automated some of the data cleaning and feature engineering processes, but now what? Your model is not creating business value unless it's deployed in production, creating insights that decision makers can take action on and incorporate into their tactical or business strategy.

But what does it mean to deploy a model to production? It depends. Are you doing an online inference or batch inference? Is there a requirement on latency or how fast your model has to make predictions? Will all the features be available at prediction time, or will we have to do some preprocessing to generate features on the fly?

Typically infrastructure is involved; either some kind of container registry or orchestration tool is used, but we might also have caches to support low-latency workflows, GPUs to speed up tensor operations during prediction, or have to use APIs if we deploy an online solution. We'll cover the basics of infrastructure and even show how you can deliver continuous value from your models through continuous integration and delivery pipelines.

Monitoring Models in Production

Okay, you've deployed your model to production. Now what? You have to monitor it. You need a way to peer underneath the covers and see what is happening in case something goes wrong. As data scientists we are trained to think of model accuracy and maybe have an awareness of how efficient one model is compared with another, but when using cloud services, we must have logging and exception handling for when things go wrong. Again, Murphy's Law is a guiding principle here.

Understanding the value of setting up logging and explicit exception handling will be a lifesaver when your model chokes on data in production it has never seen before. In subsequent chapters you'll learn to think like an engineer to add logging to your models and recover gracefully in production.

Example Components of a Production Machine Learning System

A production machine learning system has many supporting components that go into its design or *technical architecture*. You can think of the technical architecture as a blueprint for creating the entire machine learning system and may include cloud storage accounts, relational and nonrelational (NoSQL) databases, pipelines for training and prediction, infrastructure for reporting to support automated training, and many other components. A list of some of the components that go into creating a technical architecture include

- **Cloud storage accounts**

- **Relational or NoSQL databases**

- **Prediction pipeline**

- **Training pipeline**

- **Orchestration pipelines**

- **Containers and container registries**

- **Python packages**

- **Dedicated servers for training or model serving**

- **Monitoring and alerting services**

- **Key Vault for secure storage of credentials**

- **Reporting infrastructure**

Measuring the Quality of Data Science Projects

The goal of this section is to give you the ability to quantitatively define and measure and evaluate the success of your own data science projects. You should begin by asking what success means for your project.

This quantitative toolbox, akin to a kind of multi-dimensional measuring tape, can be applied to many types of projects from traditional supervised, unsupervised, or semi-supervised machine learning projects to more custom projects that involve productionizing ad hoc data science workflows.

Measuring Quality in Data Science Projects

With the rapid evolution of data science, a need has arisen for MLOps, which we discussed, in order to make the process effective and reproducible in a way that mirrors scientific rigor.

Measuring software project velocity and other KPIs common to project management, an analogous measurement is needed for data science. Table 1-2 lists some measurements that you might be interested in tracking for your own project. In later chapters we'll show you how to track these or similar measures in production using tools like MLflow so that you can learn to move the needle forward.

Table 1-2. *Common KPIs*

Measurement	Stakeholder Question	Examples
Model accuracy	Can we evaluate model performance?	Precision, recall, accuracy, F1 score; depends on the problem and data
Model interpretability	How did each feature in the model contribute to our prediction?	Shap values
Fairness	Are there sensitive features being used as input into the model?	Model output distribution
Model availability	Does the model need to make predictions at any time of day or on demand? What happens if there is downtime?	Uptime
Model reliability	Do the training and inference pipelines fail gracefully? How do they handle data that has never been seen before?	Test coverage percentage
Data drift	What happens when features change over time?	KL divergence
Model drift	Has the business problem changed?	Distribution of output of model
Correlation with key KPIs	How do the features and prediction relate to key KPIs? Does the prediction drive key KPIs?	Correlation with increased patient hospital visits for a healthcare model
Data volume	What is the size of our data set?	Number of rows in feature sets

(continued)

Table 1-2. (*continued*)

Measurement	Stakeholder Question	Examples
Feature profile	What kinds of features and how many?	Number of features by category
Prediction latency	How long does the user have to wait for a prediction?	Average number of milliseconds required to create features at prediction time

Importance of Measurement in MLOps

How can we define the success of our projects? We know intuitively that the code we build should be reliable, maintainable, and fail gracefully when something goes wrong, but what is reliability and maintainability? We need to take a look at each of these concepts and understand what each means in the context of data science.

What Is Reliability?

Reliability means there are checks, balances, and processes in place to recover when disaster strikes. While availability is more about uptime (is the system always available to make a prediction when the user needs it?), reliability is more about acknowledging that the system will not operate under ideal conditions all the time and there will be situations that are unanticipated. Since we cannot anticipate how the system will react when faced with data it's never seen before, for example, we need to program in error handling, logging, and ensure we have proper tests in place to cover all code paths that could lead to failure. A cloud logging framework and explicit exception handling are two ways to make the system more reliable, but true reliability comes from having redundancy in the system, for example, if you're building an API for your model, you should consider a load balancer and multiple workers.

What Is Maintainability?

Maintainability is related to code quality, modularity, and readability of the code. It requires a future-oriented mindset since you should be thinking about how you will maintain the code in the future. You may have an exceptional memory, but will you be able to remember every detail a year from now when you have ten other projects? It's best to instead focus on maintainability early on so running the project in the future is easier and less stressful.

Moving the Needle: From Measurement to Actionable Business Insights

Ultimately the goal of MLOps is to move the needle forward. If the goal of the data scientist is to create accurate models, it's the job of the machine learning engineer to figure out how to increase accuracy, increase performance, and move business KPIs forward by developing tools, processes, and technologies that support actionable decision making.

Hackerlytics: The Mindset of an MLOps Role

Finally, I would like to close out this chapter by discussing the mindset shift required for an MLOps role from data scientist. If you are a data scientist focused on data and analytics, you are probably used to thinking outside the box already to solve problems. The shift to using this out-of-the-box thinking to solve data problems with technology is exactly the mindset required. In the next chapter, we'll look at the fundamental skills from mathematical statistics to computer science required so you can begin to apply your new hackerlytics skills on real data problems.

Summary

In this chapter we gave an introduction to MLOps from the data scientist's point of view. You should now have an understanding of what MLOps is and how you can leverage MLOps in your own projects to bring continuous value to stakeholders through your models. You should understand some of the technical challenges that motivate the need for an MLOps body of knowledge and be able to measure the quality of data science projects to evaluate technical gaps. Some of the majors topics covered were as follows:

- What is MLOps?

- The need for MLOps

- Measuring quality of data science projects

In the next few chapters, we will cover some core fundamentals needed for data scientists to fully take ownership of the end-to-end lifecycle of their projects. We'll present these fundamentals like algorithmic and abstract thinking in a unique way that can help in the transition from data science to MLOps.

CHAPTER 2

Foundations for MLOps Systems

"All models are wrong, but some are useful."

—George Box

In this chapter, we will discuss foundations for MLOps systems by breaking down the topic into fundamental building blocks that you will apply in future chapters. While we will discuss programming nondeterministic systems, data structures and algorithmic thinking for data science, and how to translate thoughts into executable code, the goal is not to give a fully comprehensive introduction to these areas in a single chapter but instead provide further resources to point you in the right direction and answer an important question: Why do you need to understand mathematics to develop and deploy MLOps systems?

This book would be remiss without laying out the core mathematical and computational foundations that MLOps engineers need to understand to build end to end systems. It is the responsibility of the MLOps engineer to understand each component of the system even if it appears like a "black box."

Toward this end, we will create a logistic regression model (both classical and Bayesian) from scratch piece by piece to estimate the parameters of the hypothesis using stochastic gradient descent to illustrate

© Dayne Sorvisto 2023
D. Sorvisto, *MLOps Lifecycle Toolkit*, https://doi.org/10.1007/978-1-4842-9642-4_2

how models built up from simple mathematical abstractions can have robust practical uses across various industries. First, let's define what we mean by model by taking a look at some statistical terminology.

Mathematical Thinking

Mathematics is the foundation of data science and AI. Large language models like ChatGPT are transforming our lives. Some of the first large language models such as BERT (an encoder) are based on either an encoder or decoder transformer architecture. While the attention layers that are part of this model are different both in form and in use cases, they are still governed by mathematics.

In this section, we lay out the rigorous foundations for MLOps, diving into the mathematics behind some of the models we use.

Linear Algebra

Linear algebra is the study of linear transformations like rotation. A transformation is a way of describing linear combinations of vectors. We can arrange these vectors in a matrix form, and in fact you can prove every linear transformation can be represented in this way (with respect to a certain basis of vectors). You'll find linear algebra used throughout applied mathematics since many natural phenomena can be modeled or approximated by linear transformations. The McCulloch-Pitts neuron or perceptron combines a weight vector with a feature vector using an operator called the dot product. When combined with a step activation or "threshold" function, you can build linear classifiers to solve binary classification problems.

Though matrices are two-dimensional, we can generalize the idea of a matrix to higher dimensions to create tensors. Since many machine learning algorithms can be written in terms of tensor operations. In fact tensors themselves can be described as multilinear maps[1]. You can imagine how important linear algebra is to understanding neural networks and other machine learning algorithms. Another important reason for studying linear algebra is it is often the first exposure to writing proofs, developing mathematical arguments and mathematical rigor.

Probability Distributions

By model we really mean a probability distribution. The probability distribution will be parameterized so that we can estimate it using real-world data either through algorithms like gradient descent or Bayes' rule (this may be difficult under some circumstances as we'll discuss). We're usually interested in two types of probability distributions: joint probability distributions and conditional distributions.

Join probability distribution: Given two random variables X and Y, if X and Y are defined on the space probability space, then we call the probability distribution formed by considering all possible outcomes of X and Y simultaneously the *joint probability distribution*. This probability distribution written as P(X, Y) encodes the marginal distributions P(X) and P(Y) as well as the conditional probability distributions. This is an important concept as many of the models we'll be looking at will be attempting to compute or sample from a joint probability distribution to make some prediction.

Conditional probability distribution: Conditional probability is the probability of an event, Y occurring given an event X has already occurred. We write this conditional probability as P(Y | X) often read as "probability of Y given X." Let's look at a few examples of models we might use as data scientists in various industries to understand how these abstractions are built up from mathematical concepts.

---

[1] An introduction to linear algebra can be found in Hoffman, K. A. (1961). *Linear Algebra*.

Understanding Generative and Discriminative Models

A generative model is synonymous with a joint probability distribution P(X, Y) (however, this is not strictly true since, e.g., GANs belong to the class of generative models) since for a classification problem it will assume some functional form of P(Y) and P(X | Y) in terms of some parameters and estimated from the training data. This is then used to compute P(Y | X) using Bayes' rule. These types of models have some interesting properties, for instance, you can sample from them and generate new data. Data augmentation is a growing area especially within the healthcare and pharmaceutical industry where data from clinical trials is costly or not available.

The simplest examples of a generative model include Gaussian distributions, the Bernoulli model, and Naive Bayes models (also the simplest kind of Bayesian network).

In contrast, a discriminative model such as logistic regression makes a functional assumption about the form of P(Y | X) in terms of some parameters W and b and estimates the parameters directly from the training data. Then we pick the most likely class label based on these estimates. We'll see how to compute parameters W and b in the lab: algorithmic thinking for data science[2] where we'll actually use stochastic gradient descent and build a logistic regression model from the ground up.

---

[2] For a full introduction to algorithmic thinking and computer programming, the reader is directed to Abelson, H. and Sussman, G. J. (1996). *Structure and Interpretation of Computer Programs, second edition.* MIT Press.

Bayesian Thinking

We chose logistic regression as an example in this chapter for another reason: Logistic regression is a good example of a probabilistic model. When you train it, it automatically gives you an estimate of the probability of success for new data points. However, classical logistic regression is a *frequentist* model. The classical logistic regression model does not tell us if we can rely on the results, if we have enough data for training, or anything about the certainty in the parameters.

To illustrate this point, let's suppose we train a logistic regression model to predict who should receive a loan. If our training data is imbalanced, consisting of 1000 people and 900 of which are examples of people we should not lend to, our model is going to overfit toward a lower probability of loan approval, and if we ask what is the probability of a new applicant getting a loan, the model may return a low probability. A Bayesian version of logistic regression would solve this problem. In the lab, you will solve this problem of imbalance data by using a Bayesian logistic regression model and generating a trace plot to explore the parameters and ensure that the parameters are well calibrated to the data. Figure 2-1 shows a trace plot generated from this lab.

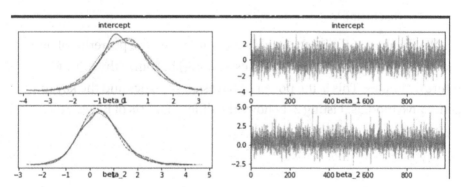

Figure 2-1. A trace plot showing the history of parameters in a Bayesian model

Of course, we need to understand yet another mathematical primitive: Bayes' rule. Unlike in frequentist statistics, where we have parameters and point estimates, in Bayesian statistics, we have probability distributions as we defined earlier. In fact, every unknown in our model is a probability distribution called a prior that encodes our current knowledge about that parameter (in the lab, we have three parameters we want to estimate, with priors chosen from normal distributions).

Bayes' rule updates beliefs about the parameters by computing a posterior probability distribution.

- The **prior** distribution can be interpreted as the current knowledge we have on each parameter (it may only be a best guess).

- The **likelihood function** is the probability of observing a data set given certain parameters θ of our model.

- The **evidence** is the probability of the observed data itself over all possible models and is very difficult to compute, often requiring multivariate integrals in three or more dimensions. Fortunately, for many problems, this is only a constant of proportionality that can be discarded[3].

We speak of "turning the Bayesian crank" when the posterior of one problem (what we are interested in estimating) becomes the prior for future estimates. This is the power of Bayesian statistics and the key to generative models. Listing 2-1 shows the different parts of Bayes' rule.

---

[3] Hoff, P. D. (2009). A First Course in Bayesian Statistical Methods. In *Springer texts in statistics*. Springer International Publishing. https://doi.org/10.1007/978-0-387-92407-6.

Listing 2-1. Bayes' rule

$$P(\theta \mid X, y) = \frac{P(y \mid X, \theta) P(\theta)}{P(y \mid X)}$$

Bayes rule was actually discovered by Thomas Bayes, an English Presbyterian minister and statistician in the eighteenth century, but the work *LII. An Essay Towards Solving a Problem in the Doctrine of Chances* wasn't published until after Bayes' death and to this day is often a graduate level course not taught in undergraduate statistics programs.

So how can we develop some intuition around Bayes' rule? Let's start by asking a question:

What is the probability of a coin coming up heads? Take a few minutes to think about it before you answer; it's a bit of a trick question.

Okay ...I've asked this question to a few people and most would say it depends. It depends if the coin is fair or not. Ok so assuming it's a fair coin, the usual answer is the chance of coming up heads is 50% or 0.5 if we're using probability.

Now let's switch this question up; let's suppose that the coin has already been flipped but you cannot see the result. What is the probability? Go ahead and ask your colleagues or friends this question, and you might be surprised by the range of answers you'll receive.

A *frequentist* position is that the coin has already been flipped, and so it is either heads or tails. The chance is either 0% heads if the coin landed tails or it is 100% if it landed heads. However, there's something unsatisfactory about this perspective; it does not take into consideration the uncertainty in the model.

A Bayesian approach would be to quantify that uncertainty and say, it's still 50% chance of heads and 50% chance of tails; it depends on *what we know* at the moment. If we observe the coin has landed heads, then we can update our hypothesis. This allows us to adapt to change and accommodate new information (remember, MLOps is all about being able to adapt to change). In the next section, we will look at some specific examples of Bayesian models.

Gaussian Mixture Models

K-means, one of the oldest clustering methods, behaves poorly when clusters are of different sizes, shapes, and densities. While K-means requires knowing the number of clusters as a parameter before going ahead with clustering, it is closely related to nonparametric Bayesian modeling in contrast to the Gaussian mixture model (GMM) shown in Figure 2-2.

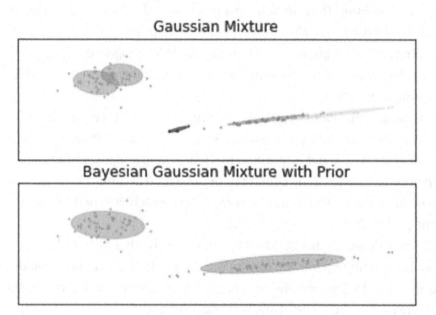

Figure 2-2. *Clustering using a Bayesian Gaussian mixture model*

A Gaussian mixture model is a probabilistic model that makes the assumption that the data generating process is a mixture of finite Gaussian distributions (one of most important probability distributions for modeling natural phenomena and so widely used in science, engineering, and medicine). It is a parametric model where the parameters of the Gaussian components are unknown. We can think of GMMs as a finite weighted sum of Gaussian component densities. Listing 2-2 shows an equation that governs the GMM.

Listing 2-2. An equation that describes the Gaussian mixture model

$$p(x) = \sum_{i=1}^{m} \theta_i \mathcal{N}\left(x \mid \mu_i, \sum_i\right)$$

However, these parameters are computationally expensive and do not scale well since they are computed often through maximal likelihood and EM (expectation-maximization) algorithms and are modeled as latent variables. Despite scalability problems, GMMs are used in many industries in particular healthcare, to model more meaningful patient groupings, diagnostics, and rehabilitation and to support other healthcare activities.

General Additive Models

With generalized additive models (GAMs), you don't have to trade off accuracy for interpretability. These are a very powerful extension to linear regression and are very flexible with their ability to incorporate nonlinear features in your data (imagine having to do this by hand if all we had was a linear regression model?)

If random forests are data driven and neural networks are model driven, the GAMs are somewhere in the middle, but compared to neural nets, SVMs, or even logistic regression, GAMs tend to have relatively low misclassification rates which make them great for mission critical applications where interpretability and misclassification rate are utmost importance such as in healthcare and financial applications.

If you've never used a GAM before, you can look at splines to start. Splines are smooth functions used to model nonlinear relationships and allow you to control the degree of smoothness through a smoothing parameter. Figure 2-3 shows some of the trade-offs between these different models.

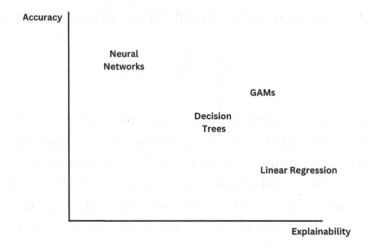

Figure 2-3. *Trade-offs between explainability and model accuracy*

Kernel Methods

The best known example of kernel methods, the support vector machine (SVM), allows us to use the "kernel trick" and work well on text based data which is naturally high dimensional. This means that we can embed features in a higher, possibly infinite, dimensional space without ever having to explicitly compute the embedding.

Kernel methods are particularly important in geostatistics applications such as kriging (Gaussian process regression) especially in the oil and gas industry. The main use case of kriging is to estimate the value of a variable over a continuous spatial field. For example, you may have sensor readings such as temperature and pressure in an oil and gas reservoir, but you may not know the sensor readings at every position in the reservoir. Kriging provides an inexpensive way to estimate the unknown sensor readings based on the readings that we do know. Kriging uses a covariance matrix and a kernel function to model the spatial relationships and spatial dependencies of data points throughout the reservoir by encoding similarity between features.

Where did these models come from? Fundamentally, these algorithms are based on logic and mathematical properties and primitives such as probability distributions. If we know about the Gaussian distribution, it's not a far stretch to understand how to build a Gaussian mixture model. If we know about covariance matrices, we can understand kernel methods and maybe Gaussian processes, and if we understand linear systems, we can build on top of this abstraction to understand GAMs.

I want to illustrate this point further, by building a model from fundamental principles. Logistic regression is particularly interesting to use for this as most people are familiar with it, but it is not a toy model; you can use logistic regression to solve many real-world problems across various industries since it is fairly robust. We can also use logistic functions to build many more complex models. For instance, the classical version of logistic regression is used to model binary classification (e.g., predicting likelihood of success or failure), but by combining multiple logistic regression models into strategies like one-vs.-all or one-vs.-one, we can solve more complex multi-class classification problems with a single response variable via softmax, a generalization of logistic regression. Logistic functions can also be used to create neural networks: It's no coincidence that in a neural network, the activation function is often a logistic sigmoid (pictured in the following). In this chapter's lab on algorithmic thinking, you're going to walk through some of these mathematical tools and build a logistic regression model from scratch. Figure 2-4 shows an example of a logistic sigmoid curve.

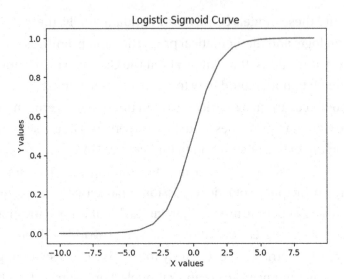

Figure 2-4. *A logistic sigmoid curve*

The parameter mu takes on the probability value ½ which makes intuitive sense.

Higher Dimensional Spaces

I want to cover one more mathematical tool for the toolkit because in data science, unlike pure statistics, we deal with big data but also because it is fascinating to be able to develop intuition on higher dimensional spaces by learning to think geometrically.

You've probably heard of the "curse of dimensionality." Things behave strangely in high dimensions, for example, if we could measure the volume of a unit sphere as we embed it into higher dimensional space, that volume would actually shrink as the dimension increases! That is incredibly counterintuitive. Figure 2-5 shows an artistic rendition of a shrinking sphere in higher dimensional space (since we can only visualize in three dimensions).

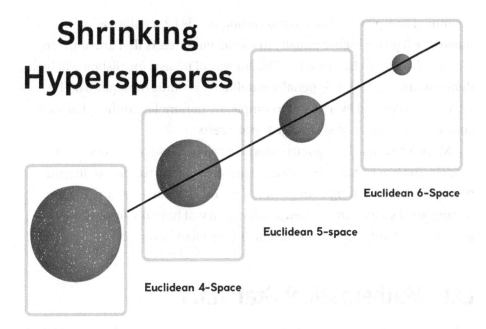

Figure 2-5. *A shrinking sphere illustrating the unintuitive nature of higher dimensions*

In data science in the real world, we have at minimum hundreds of features. It is not uncommon to have 1000 or more features, and so naturally we need a way to try to reduce the number of features. Mathematically speaking, this means we want a way to embed our data that lives in a high dimensional space to a lower dimensional space while preserving information.

Going back to our favorite example, logistic regression, we can illustrate another important mathematical tool to handle high dimensionality, regularization.

Regularization is extremely important when applying logistic regression because without it, the asymptotic nature of the logistic curve at +infinity and -infinity (remember the sigmoid?) would translate into zero loss in high dimensions. Consequently, we need strategies to dampen

47

the model complexity. The most common way is L2 regularization which means we'll give a higher penalty to model parameters that are nonzero. We can also use an L1 norm (a different way of measuring distance in high dimensional spaces). The penalty is defined as minus the square of the L2 norm multiplied by a positive complexity parameter lambda. Lambda controls the amount of shrinkage toward zero.

Models that use L1 regularization are called Lasso regression, and models that use L2 are called Ridge regression. If you would like to gain a deeper understanding of the types of norms that can exist and higher dimensional spaces, in the next section, you will have the opportunity to learn more about mathematical statistics in a hands-on lab.

Lab: Mathematical Statistics

Before proceeding to the next section, you can complete the optional lab on mathematical statistics. This will give you hands-on experience with probability distributions by looking at an important and fundamental tool in mathematical statistics: characteristic functions.

You'll program a characteristic function from scratch. Characteristic functions have many interesting properties including completely characterizing a probability distribution and are even used in the most basic proofs of the central limit theorem. The steps are as follows:

Step 1. Open the notebook MLOps_Lifecycle_Toolkit_Mathematical_ Statistics_Lab.ipynb (available at github.com/apress/mlops-lifecycle-toolkit).

Step 2. Import the math, randon, and numpy packages by running cell #2.

Step 3. Create a function for computing the characteristic function of a random normal with unit standard deviation by running cell #3.

Step 4. Run the remaining cells, to set up a coin toss experiment and recover the probability of a fair coin from the characteristic function. Was the coin fair?

Although this lab is optional because it requires some advanced math, it's recommended since it covers some deep mathematical territory from probability distributions, Fourier transforms, complex numbers, and more.

Programming Nondeterministic Systems

In order to build real-world systems, we need to understand the types of data structures (arrays, lists, tensors, dataframes) and programming primitives (variables, loops, control flow, functions) that you'll likely encounter to know what the programming is doing and to be able to read other data scientists code.

Knowledge of data structures, algorithms, and packages can be applied regardless of language. If you use a package, even an R package, you should read the source code and understand what it's doing. The danger of not understanding what the statistical black box means the result of an analysis that uses your code could come out inaccurate or, worse, introduce non-determinism into your program.

Sources of non-determinism in ML systems

- *Noisy data sets*

- *Poor random initialization of model parameters*

- *Black box stochastic operations*

- *Random shuffling, splits, or data augmentation*

Programming and Computational Concepts

Let's look at some basic programming concepts.

Loops

Loops are a mechanism to repeat a block of code. Why use loops? One reason is you may have a block of code you want to repeat and, without a loop, you would have to copy paste the code, creating redundant code that is hard to read and reason about.

Another reason we use loops is for traversing a data structure such as a list or dataframe. A list or array has many items and a dataframe has many rows, in a well-defined order, and it is a natural way to process each element one by one; whether that element be a row or a list depends on the data structure.

Loops can be complex, and there's a programming adage that goes you should never modify a variable in a loop.

One reason loops are important in data science is twofold:

1) Many tensor operations naturally unfold into loops (think dot product or tensor operations).

2) By counting the number of nested loops, you can get an idea on the asymptotic behavior (written in Big-O notation) of your algorithm; in general, nested loops should be avoided if possible being replaced by tensor operations.

The last technique is actually an optimization tool called **vectorization**. Often, vectorized code can take advantage of low level instructions like single instruction, multiple data, or SIMD instructions. In fact, most GPUs use a SIMD architecture, and libraries like JAX can take this idea to the next level if you need to run NumPy code on a CPU, GPU, or even a TPU for high performance machine learning.

Variables, Statements, and Mathematica Expressions

What is the difference between a statement and an expression?

A statement does something that assigns a value to a variable. An example in Python is x = 1.

This simple statement assigns the value 1 to a variable x. The variable, x, in this case points to a memory location used to store information.

An expression on the other hand needs to be evaluated by the interpreter (or compiler in a compiled language like C++ or Haskell) and returns a value. Expressions can be building blocks of statements or complex mathematical expressions. An example of an expression (but not a statement) is the following:

```
(1 + 2 + x)
```

We can also have Boolean expressions which we'll look at next and are very important for making decisions.

Control Flow and Boolean Expressions

Control flow refers to the order in which individual statements, commands, instructions, statements, or function calls are executed. Changing the order of statements or function calls in a program can change the program entirely. In imperative languages (e.g., Python can be coded in an imperative style), control flow is handled explicitly by control flow statements such as if statements that control branching. Usually at each branch, a choice is made and the program follows one path depending on a condition. These conditions are called Boolean expressions.

Boolean expressions involve logical operations such as AND, OR, NOT, and XOR. These Boolean expressions can be combined in complex ways using parentheses and as mentioned are used in control flow statements in your program to make complex decisions.

For example, let's suppose you have a computer program with variables that store true and false values. You have one variable that stores the percent missing and a second variable that stores the number of rows in your data, and you want to exclude rows that have over 25% missing values when your data is more than 1000 rows. You can form a Boolean expression as follows:

```
If (percent_missing > 25) AND (num_rows  >  1000):
        // drop rows
```

Of course, in a library like Pandas, there are functions like dropna for dataframes that do this sort of low level logic for you, but you can read the source code to understand exactly what is happening under the hood for the functions you care about.

Tensor Operations and Einsums

A tensor is, simply put, a generalization of vectors to higher dimensions. There is some confusion on the use of the term since there are also tensors in physics, but in machine learning, they're basically a bucket for your data. Many libraries including NumPy, TensorFlow, and PyTorch have ways of defining and processing tensors, and if you've done any deep learning you're likely very familiar with tensors, but a cool tool I want to add to your toolkit is Einsums.

Einsums are essentially shorthand for working with tensors, and if you need to quickly translate complex mathematical equations (e.g., ones that occur in data science or machine learning papers), you can often rewrite them in Einsum notation in very succinct, elegant ways and then execute them immediately in a library like PyTorch. For example, the following Einsum equation codifies matrix multiplication, and we can implement it in PyTorch in Listing 2-3:

Listing 2-3. An example of Einsum notation

```
a = torch.arange(900).reshape(30, 30)
b = torch.arange(900).reshape(30, 30)
torch.einsum('ik,kj->ij', [a, b])
```

Okay, we've covered quite a bit. We talked about variables, loops, and control flow and ended with tensors, a kind of bucket for high dimensional data. However, there are many more "buckets" for your data that are useful in data science. These are called data structures, the subject of computer science. We'll cover a few data structures in the next section.

Data Structures for Data Science

This section is about data structures. While computer science has many data structures, data scientists should be familiar with a few core data structures like sets, arrays, and lists. We will start by introducing sets, which might be the simplest data structure to understand if you come from a math background.

Sets

Sets are collections of elements. A set can contain elements, and you can use sets for a variety of purposes in data science for de-duplication of your data to checking set membership (that is to say, the set data structure comes with an IN operator).

It is important to note that a set has no order (actually there is the well-ordering principle that says exactly the opposite, but in Python, for instance, and other languages, sets have no order). If we want to impose an order when storing elements, we should use a linear data structure like an array or a list, which we'll cover next.

Arrays and Lists

The most fundamental distinction between an array and a list is that a list is a heterogeneous data structure, and this mean it can store a mix of data types, for example strings, floats, Booleans, or even more complex user defined types.

An array on the other hand is homogenous; it only is designed to store one type of value.

In Python, lists are a primitive data type and part of the core language. The ability to use list comprehensions instead of loops for mathematical constructs is very useful in data science. However, for efficient processing of data, we can use a library like NumPy which has a concept of arrays. This is known as a trade-off, and in this case, the trade-off exists between efficiency and convenience.

Part of being a good technical decision-maker is understanding these types of technical trade-offs and the consequences on your own project. For example, if you decide to profile your code and find you're running into memory errors, you might consider changing to a more efficient data structure like a NumPy array, maybe even with a 32 bit float if you don't need the extra precision of a 64 bit floating point number.

There are many different types of data structures and we'll provide resources for learning about more advanced types (one of the core subjects of computer science), but for now, we'll take a look at a more complex type that you should be aware of such as hash maps, trees, and graphs.

Hash Maps

Hash maps are an associative data structure; they allow the programmer to associate a key with a value.

They provide very fast lookup by keys, allowing you to retrieve a value corresponding to a key in O(1) time by using dynamically sized arrays under the hood and allow you to retrieve a value you've associated with your key.

If you didn't have this kind of associative data structure, you'd have to, for instance, store your elements as an array of tuples and would need to write code to search for each key you wanted to locate in the array. This would not be very efficient, so when we want to associate one piece of information with another and only care about being able to retrieve the value we've mapped to a particular key, we should consider hash maps. The point is, having a command of data structures can simplify your code drastically and make it more efficient.

In Python, a hash map is called a dictionary. One point to keep in mind when using hash maps is that the keys should be hashable, meaning a string is OK for a key but a mutable data type like a list that can be changed is not allowed.

Trees and Graphs

A graph is a mathematical data structure consisting of nodes and edges. The nodes are also called vertices. The difference between a tree and a graph is that a tree has a root node. In a graph there is no root node that is unique but both structures can be used for representing many different types of problems. Graph neural networks and graph databases are huge topics today in machine learning and MLOps, and part of the reason is that a graph, like a set, is a very general mathematical way of representing relationships between concepts that can be easily stored on a computer and processed.

You should be aware of a couple kinds of trees and graphs in particular binary trees and DAGs.

Binary Tree

A binary tree is a tree (it has a root node), and each node including the root has either 2 (hence binary) children or 0 children (in this case, we call it a leaf node). A picture of a binary tree is shown in Figure 2-6.

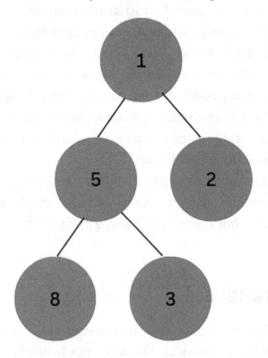

Figure 2-6. *A binary tree*

Binary trees can be complete or perfect or have additional structure that makes them useful for searching such as binary search trees.

DAGs

A graph is a generalization of a tree. A graph however can have cycles, meaning if you were to visit every node and follow its neighbor, you may find yourself in an infinite loop. An important type of graph with no cycles is called an acyclic graph and is often used in MLOps tools like Airflow to represent data flow. Directed acyclic graphs are called "DAGs" and have a variety of uses in MLOps (e.g., the popular Airflow library uses DAGs for creating pipelines).

SQL Basics

We've covered programming languages like Python, but you also need to know how to manipulate data in your programs. SQL is actually based on relational algebra and the set data structure we covered previously (the foundations were written by Edger F. Codd). SQL consists of queries and the queries can be broken down into statements. A SQL statement consists of the following clauses executed in the following order:

- FROM

- JOINS on other tables

- WHERE clause for filtering data

- GROUP BY for aggregating by multiple columns

- HAVING for filtering after aggregation

- SELECT for selecting columns or fields you want to use in your data set

- ORDER BY for sorting data by one or more columns (this can cause performance issues and should only be used sparingly)

A common table expression or CTE is a very useful construct when operationalizing data science code. The reason it is so powerful is that a CTE allows you to think algorithmically, by breaking down your SQL query into a series of steps. Each step can depend on previous steps and is materialized as a kind of "virtual table." A simple example of a CTE is given in the following; this CTE first creates a base table called Sensor_CTE and then selects from it in Listing 2-4.

Listing 2-4. An example of a common table expression or CTE

```
-- An example of a CTE
WITH Sensor_CTE (SalesPersonID, SalesOrderID, SalesYear)
AS
-- Define the CTE query.
(
    SELECT ID as Component, MAX(Pressure) as Pressure,
    AVG(Temperature) as Temperature
    FROM Sensor.Readings
    WHERE ID IS NOT NULL
    GROUP BY ID
)
-- Define the outer query referencing the CTE name.
SELECT Component, Temperature
FROM Sensor_CTE;
```

Understanding how joins and common table expressions (CTEs) work is typically what separates beginners from advanced SQL users. Most data science code requires multiple passes on data sets, and CTEs are a natural way to write more complex SQL code that requires multiple steps to process data.

Algorithmic Thinking for Data Science

An algorithm is essentially a set of rules or instructions for performing calculations that occur in a certain sequence. You can think of it like a recipe. Unlike recipes though, algorithms will usually involve data structures for storing data, and the heart of the algorithm will be manipulating these data structures to solve a problem. Unfortunately, you need to learn algorithmic thinking, and by doing so, we've created a lab for you. In the lab, you're going to start with data structures we've learned to build some basic mathematical primitives like sigmoid function and logistic curve and combine these abstractions to build your own logistic regression model. Refer to the Jupyter notebook labs for this chapter entitled "Building a Logistic Regression Model from Scratch," and complete the lab before continuing to the next section.

Core Technical Decision-Making: Choosing the Right Tool

Beyond this section, we're going to assume you've completed the labs and have a basic grasp on programming fundamentals. Before covering specific packages and frameworks for translating experiments and thoughts into executable code, I want to discuss technical decision-making briefly and how we should think about choosing the right framework for our problem.

The most important criterion in the real world is considering what tools and frameworks are already being used by your organization, colleagues, and the community behind the framework. Although you might be tempted to use a package from a language like Julia or Haskell, you should carefully consider whether or not you'll have to translate your problem into another language at some point in the future if either the package is no longer supported or because nobody in your organization has the skill set required.

Translating Thoughts into Executable Code

You might want to choose one of the following packages and dive deeper into some frameworks that are used in the real world to build machine learning models. In later chapters, we'll walk you through how to create your own packages. The important thing here is understanding that these tools we depend on in data science like Pandas or Numpy or PyTorch are just packages someone (or a team of people) have written and created. You too can learn to create your own packages, but first we need to understand why we use packages and how it makes our lives as both data scientists and MLOps engineers easier.

Understanding Libraries and Packages

What is the point of a software package? Why not use a notebook? Packages allow us to bundle code together and give it a name, import it, and reference objects inside the package so we can reuse them without having to rewrite those objects. Packages can also be versioned (see semantic versioning[4]).

For example, you may have heard of RStudio package manager for R or pip for Python. Before experimenting with any of the packages listed in the following, you should understand the package manager in your language of choice so you can install the package. We also recommend environments to isolate dependencies. We'll cover the gritty details of package managers and environments in Chapter 3, but for now here is a broad overview of some of the most interesting packages you might come across as an MLOps engineer.

---

[4]Semantic versioning 2.0.0 can be found at `https://semver.org/`.

PyMc3 Package

An active area of research is in probabilistic programming. The PyMc3 library contains various primitives for creating and working with random variables and models. You can perform MCMC (Markov chain Monte Carlo) sampling and directly translate statistical models into code.

Something to keep in mind is at the current time, these algorithms may not be very scalable, so you'll usually only see Bayesian optimization applied to the hyperparameter search part of a machine learning lifecycle using libraries like HyperOpt; however, we mention probabilistic programming as Bayesian statistics is slowly becoming a part of mainstream data science.

Numpy and Pandas

Numpy and Pandas are the bread and butter of most data science workflows. We could write an entire chapter covering just these libraries, but we'll mention for the uninitiated that Pandas is a data wrangling library. It provides a data structure called a DataFrame for processing structured data and various methods for reading csv files and manipulating dataframes. NumPy has the concept of ndarrays and allows you to process numerical data very fast without having to know much about C++ or low level hardware.

R Packages

R uses a package system called CRAN which makes available R binaries. Unlike Python, CRAN packages typically have higher dependency on other packages and tend to be focused on specific areas of statistical computing and data visualization.

The reason data scientists still use R is many packages written by researchers and statisticians are written in R. However, you should be aware of the following interoperability and scalability issues with R:

- R is not as widely supported; for example, the machine learning SDK uses R, but there is a lag between when features are released in Python and when they become available in R.

- Writing clear, concise and easy to read code in R requires considerable skill and even then there are leaky abstractions which make code difficult to maintain such as

- R is not scalable and has memory limitations. For scalable R, we recommend Databricks using SparkR.

A lot of R packages revolve around the TidyVerse. You should be familiar with the following basic R packages:

Deplyr: Deplyr is a package that is similar to Pandas in Python and is used for data wrangling. The package provides primitives such as filter and melt.

Shiny: The R ecosystem's answer to dashboarding in data science, Shiny is a package for authoring dashboards in R and fulfills the same need as Dash in Python. The advantage of ShinyR is you can build web apps without having to know how web development works. The web apps can be interactive, and you can interact with different panels of the dashboard and have multiple data sources to visualize data sets. We don't recommend Shiny as it can be hard to deploy to a web server securely.

SAS: SAS is a language of statistical programming. SAS is a procedural language. SAS requires a SAS license and is common in healthcare and finance industry where exact statistical procedures need to be executed.

MATLAB/OCTAVE: MATLAB and the open source version Octave are libraries for linear algebra. If you are prototyping a machine learning algorithm whose primitives can be expressed using matrix operations (which is a lot of machine learning), then you might consider using one of these languages. MATLAB is also particularly popular in engineering disciplines for simulations and is used in numerical computing.

PySpark: Spark is a framework for distributed computing and has a tool called PySpark that allows you to write code similar to Pandas using dataframes but in a scalable way. You can translate between Pandas and Pyspark using the latest Pandas API for spark (replacement for Koalas) and process gigabytes or even terabytes of data without running into out of memory errors. Other alternatives are called "out of core" solutions and include Dask or Modin that utilize disk storage as an extension of core memory in order to handle memory-intensive workloads.

Important Frameworks for Deep Learning

There are many frameworks in Python for deep learning and working with tensors. PyTorch and TensorFlow 2.0 with Keras API are the most popular. Although we could implement our own routines in a package like the NumPy example to build your own 2D convolutional layer and use these functions to build a convolutional neural network, in reality, this would be too slow. We would have to implement our own gradient descent algorithm, auto differentiation, and GPU and hardware acceleration routines. Instead, we should choose PyTorch or TensorFlow.

TensorFlow

TensorFlow is an end to end machine learning framework for deep learning. TensorFlow is free and open sourced under Apache License 2.0 and supports a wide variety of platforms including MacOS, Windows, Linux, and even Android. TensorFlow 1.0 and TensorFlow 2.0 have significant differences in APIs, but both provide the tensor as a core abstraction allowing the programmer to build computational graphs to represent machine learning algorithms.

PyTorch

The advantage is PyTorch is class oriented, and if you have a strong Python background, you can write a lot of custom code in an object oriented style without having to be very familiar with how the APIs work like in TensorFlow. PyTorch for this reason is used in academic papers on machine learning and is a solid choice for prototyping machine learning solutions.

Theano

PyMC3 is written on top of Theano as well as some other interesting projects, but Theano is no longer supported so it is not recommended for ML development or MLOps.

Keras

Prior to the introduction of the Keras API, developers required specific knowledge of the API. Keras however is very beginner friendly, and some useful features of TensorFlow are GPU awareness (you do not need to change your code to use a GPU if one is available, as TensorFlow will detect if for you); the Keras API is very intuitive for beginners, and there is a large community around TensorFlow so bugs and CVEs (security vulnerabilities) are patched regularly. Post TensorFlow 2.0 release, you can also do dynamic execution graphs.

Further Resources in Computer Science Foundations

We've covered a lot of ground, discussed data structures and algorithmic thinking, and covered the basics of computer science required to work with data such as graphs, dataframes, tables, and the basics of SQL. We've talked about R and Python, two common languages for data science, and some of their common packages.

However, it is important to stress this is only the minimum. It would not be possible to cover a complete course in computer science for data scientists in this chapter, and so the best we can do is provide some recommended reading so you can educate yourself on topics you're interested in or fill in gaps in your knowledge to become better programmers. We've curated the following list of books on computer science that we think would be most valuable for data scientists.

- Introduction to Algorithms by Rivest[5]

- Bayesian Methods for Hackers by Davidson Pilon[6]

In general, you can read a book on functional analysis (for infinite dimensions) or linear algebra (for finite dimensional spaces) provided in the following.

Further Reading in Mathematical Foundations

Although we covered some mathematical concepts in this chapter, it would not be possible to cover even the simplest areas like linear algebra in detail without further resources. Some areas you may be interested in pursuing on your own are Bayesian statistics[7] (understanding Bayes' rule, Bayesian

---

[5] Cormen, T. H., Leiserson, C. E., Rivest, R. L., & Stein, C. (2009). *Introduction to Algorithms*. MIT Press.

[6] Davidson-Pilon, C. (2015). *Bayesian Methods for Hackers: Probabilistic Programming and Bayesian Inference*. Addison-Wesley Professional.

[7] McElreath, R. (2015). *Statistical Rethinking: A Bayesian Course With Examples in R and Stan*. Chapman & Hall/CRC.

inference, and statistical thinking), statistical learning theory[8] (the rigorous foundations of the many learning algorithms we use in MLOps), and of course linear algebra[9] (in particular finite dimensional vector spaces are a good stepping stone to understand more advanced concepts).

Summary

In this chapter, we discussed the importance of understanding mathematical concepts and how MLOps systems can be viewed as stochastic systems that are governed by mathematical abstractions. By understanding these mathematical abstractions and having an understanding of data structures and algorithmic thinking, we can become better technical decision-makers. Some of the topics we covered in this chapter include the following:

- Programming Nondeterministic systems

- Data Structures for Data Science

- Algorithmic Thinking for Data Science

- Translating Thoughts into Executable Code

- Further Resources on Computer Science

In the next chapter, we will take a more pragmatic perspective and look at how we can use these abstractions as tools and software packages when developing stochastic systems in the real world.

---

[8] Hastie, T., Tibshirani, R., & Friedman, J. (2013). *The Elements of Statistical Learning: Data Mining, Inference, and Prediction*. Springer Science & Business Media.

[9] Halmos, P. (1993). *Finite-Dimensional Vector Spaces*. Springer.

CHAPTER 3

Tools for Data Science Developers

"Data! Data Data! I can't make bricks without clay!"

—Sir Arthur Conan Doyle

How do we manage data and models? What are the tools we can use to make ourselves more efficient and agile in data science? In this chapter, we will deep dive into the tools and technology that you will depend on daily as an MLOps engineer.

AI tools can make you more productive. With the release of GPT3 in June 2020, the large language model and "brains" behind the ChatGPT app, and in March of 2023, GPT4, the first multimodal large language model capable of understanding both text and images was released. Data scientists will increasingly use AI tools to write code.

The growth is exponential, and although it cannot predict very far into the future what specific tools will be available, it is certain that basic tools like code version control systems, data version control, code editors, and notebooks will continue to be used in some form or another in data science, and what's important is to have a solid foundation in the basics.

You will understand version control, data version control, and specific python packages used at various stages of the spiral MLOps lifecycle. You should be comfortable enough at the end of this chapter to complete the

© Dayne Sorvisto 2023
D. Sorvisto, *MLOps Lifecycle Toolkit*, https://doi.org/10.1007/978-1-4842-9642-4_3

titular MLOps toolkit lab work where you'll build a cookie cutter MLOps template you can apply to accelerate your projects and be able to install a wide range of MLOps packages like MLFlow and Pandas to support various stages of the MLOpS lifecycle.

Data and Code Version Control Systems

Data science is a collaborative activity. When you are first learning data science you might spend most of your time alone, exploring data sets you choose and applying whatever models perform best on your data set.

In the real world you typically work on a team of data scientists, and even if you are the sole individual contributor on your team, you still likely report results to stakeholders, product managers, business analysts, and others and are responsible for handling changes.

All of these changes in a business impact data science as they result in changes in downstream feature engineering libraries and training scripts. You then need a way to share code snippets and get feedback in order to iterate on results and keep track of different versions of training scripts, code, and notebooks. Are there any tools to manage this change in data, code, and models? The answer is version control.

What Is Version Control?

Version control is a software tool used to manage changes in source code. The tool keeps track of changes you make to the source code and previous versions and allows you to roll back to a previous version, prevent lost work, and pinpoint where exactly in the code base a particular line was changed. If you use it properly, you read the change log and understand the entire history of your project. Git, a distributed version control system (as opposed to centralized version control), is a standard for data science teams.

What Is Git?

As we mentioned, Git is a standard tool for version control. Git works basically by taking a snapshot of each file in your directory and storing this information in an index. Git is also a distributed version control system (as opposed to a central version control like TFS) which means it supports collaboration among data scientists and developers. Each developer can store the entire history of the project locally, and because Git only uses deltas, when you are ready to commit changes, you can push them to the remote Git server, effectively publishing your changes.

Git Internals

Git uses commands. There are several Git commands you should be aware of and some special terminology like "repos" which refers to a collection of files that are source controlled. If you are unfamiliar with the concept of repos, you could think of it like a kind of directory where your source code lives.

In practice, when working on a data science project as in MLOps role, you will probably use a source control tool like Sourcetree since productivity is important, and also once you know the basics of the commands, it gets very repetitive to type each time. Tools like Sourcetree abstract these details away from you. You may be wondering why a tool like Sourcetree could help data scientists when you can use the Git command. As we will see in the next section, Git does provide low level commands for interacting with Git repositories, but Sourcetree is a GUI tool, and since typing the same command over and over again takes time, using a GUI tool will make you a more productive developer.

Plumbing and Porcelain: Understanding Git Terminology

Porcelain commands refer to high level Git commands that you will use often as part of your workflow. Plumbing is what Git does behind the scenes. An example of a porcelain command is the Git status to check for changes in your working directory.

Ref: A ref is essentially a pointer to a commit. A pointer is how Git represents branches internally.

Branch: A branch is similar to a ref in that it's a pointer to a commit. You can create a new branch using the command given in Listing 3-1.

Listing 3-1. Git command to create a new branch

```
git branch <branch>
```

Head: Some documentation makes this seem really complicated but it is not. In Git, there was a design choice that only one branch can be checked out at a time, and this had to be stored in some reference or pointer. (If you don't know what a reference or a pointer is, read the Git documentation[1]).

How Git Stores Snapshots Internally

Git assigns a special name to this pointer called HEAD. There can only be one HEAD at a time and it points to the current branch. This is the reason why you might hear HEAD referred to as the "active" branch.

You might be wondering, how is this "pointer" physically stored on a computer. Well, it turns out the pointer is not a memory address but a file. This file stores the information that the HEAD is the current branch

---

[1] The Git documentation covers topics including references and pointers: https://git-scm.com/book/en/v2/Git-Internals-Git-References

(remember the definition of a branch from earlier). There is a physical location on the computer in the .git/HEAD directory where this file is located and you can open it up in a text editor (such as Notepad++) and read its contents for yourself to understand how Git stores information internally.

Don't worry if this seems complicated, as it will be much easier in the lab work and begin to make sense when you use it and see the purpose of Git for yourself.

Sourcetree for the Data Scientist

We recommend using Sourcetree, a free open source GUI based tool. If you are a professional software developer, you can try Kraken which has some additional features but requires a license. There are two steps for using Sourcetree:

You can download Sourcetree at sourcetreeapp.com. You need to agree to terms and conditions and then download the app (Figure 3-1):

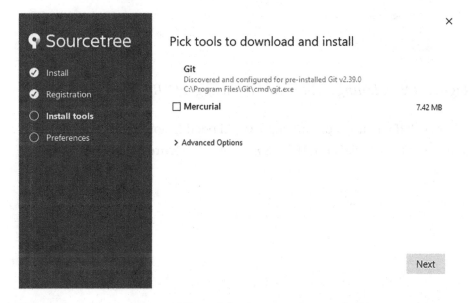

Figure 3-1. *Sourcetree GUI tool for interacting with Git repositories*

Step 1: Clone a remote repository. Figure 3-2 shows the GUI interface for cloning a repository in Sourcetree.

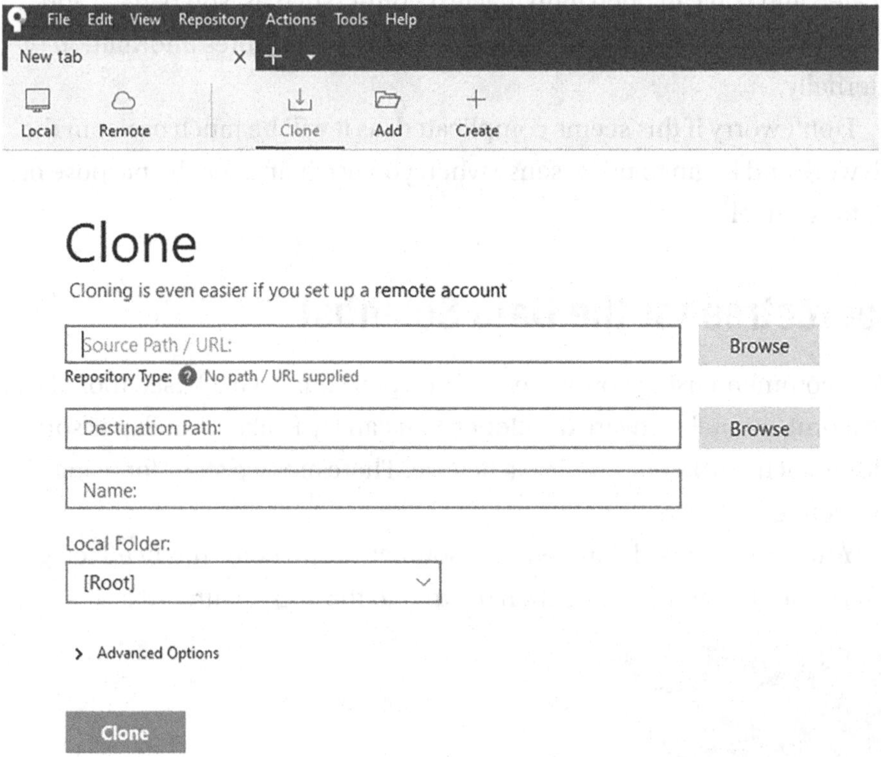

Figure 3-2. Cloning a Git repository using a GUI

Step 2: If you use a private repo, you'll need to configure your ssh key. Make sure to click SSH not HTTPS as shown in Figure 3-3.

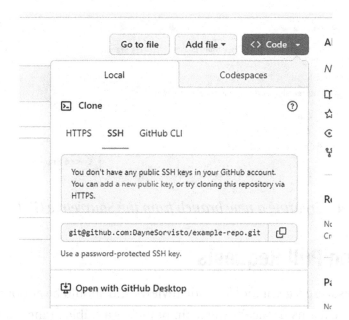

Figure 3-3. *Copying the SSH path to your GitHub repo*

Branching Strategy for Data Science Teams

If you are on a team of at least five developers, you may have to consider a branching strategy. This matters less if you are on a small team or alone because as a data scientist you may be OK to rely on a single main branch, but with more than five developers, you may consider setting up a second branch.

If you want to learn about more complex branching strategies beyond feature branches, you can read about Git Flow. Usually different branching strategies are chosen in consideration of a software release schedule in collaboration with other teams depending on the size of your organization among other factors. Figure 3-4 shows how to create a new branch from the Sourcetree GUI.

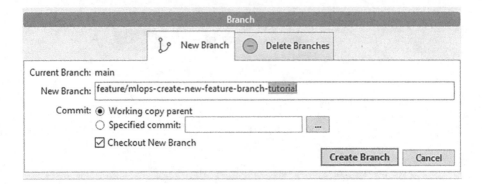

Figure 3-4. *Creating a new branch from the Sourcetree GUI*

Creating Pull Requests

Pull requests are a great tool for code reviews and should be adopted by data science teams. Typically the main branch is a stable branch, and prior to merging changes into main, you should have a peer review your changes. Ideally, a data scientist on your team that is familiar with Git would be designed as the release manager and would coordinate this with the team, but the process can be done informally. Figure 3-5 shows how to create a pull request.

Benefits of pull requests for data scientists include the following:

- Opportunity to review changes and learn new data science techniques.

- Catch mistakes and bugs before they are committed to main branch, increasing code quality metrics.

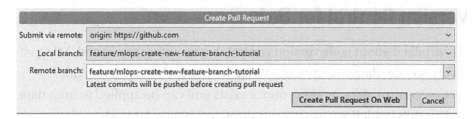

Figure 3-5. Creating a pull request

Do I Need to Use Source Control?

You might wonder if all of this is necessary or do you even need source control. But what are the consequences of not using it? You should use source control for the following reasons

- You are part of a team of data scientists sharing code and collaborating and need to coordinate changes through a remote branch.

- You want to version your data so you can train models on a previous version of data.

- You are a data scientist that does not want to lose their work on their local system and wants something more reliable than a Jupyter Notebook's autosave functionality.

- You need a way to save different snapshots of your data or code.

- You want a log or paper trail of your work in case something breaks (you can use the "Blame" feature to pinpoint the author of a change).

Version Control for Data

We've talked about code version control, but as we've mentioned, MLOps involves code, data, and models. While we use tools like Git for code version control, data version control exists and can be applied to both data and models (which are usually serialized in a binary format).

The standard package is called DVC (you can guess this stands for data version control). DVC works on top of Git, and many of the commands and terminology are similar. For example, the dvc init command is used to initialize data version control in your repo. In the lab, you'll work through some basic dvc commands for data and model version control.

Git and DVC Lab

In this lab (Figure 3-6), you will gain some hands-on experience with both Git for interacting with Git repositories and DVC for versioning data. Fortunately, many of the Git commands you will learn in the lab are very similar to the DVC commands you will learn later. However, throughout the lab, you should keep in mind the distinct purpose of each tool and where you might want to use each in your own workflow.

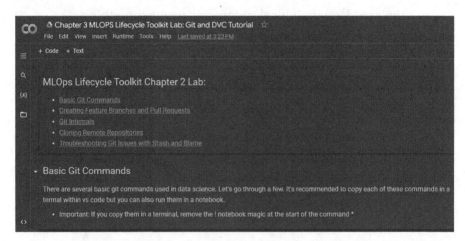

Figure 3-6. *GIT and data version control (DVC) lab*

Before proceeding to the next section on code editors, complete the version control lab titled: Chapter_3_MLOPS_Lifecycle_Toolkit_Lab_Git_and_Dvc

Step 1. Open Chapter_3_MLOPS_Lifecycle_Toolkit_Lab_Git_and_Dvc.ipynb and read the instructions.

Step 2. Copy paste the commands in the notebook into a terminal, and get familiar with each command and what it does; you can use the -h flag to see what each command does (e.g., git status -h).

Step 3. Sign up for a GitHub account by following instructions in the lab.

Model Development and Training

So we've covered version control systems for both code and data but how about the tools we use to edit our code and develop models? You may be using a tool like Spyder or a Jupyter notebook to edit your code, and surely like most developers, this is your favorite editor. I don't want to change your mind, but it's worth knowing the range of code editors available in data science and when and why you might want to consider using an editor like VS Code over Spyder.

Spyder

Spyder is a free and open scientific environment for data science. It was first released in 2009 and is available cross-platform (Windows, Linux, and MacOS) through Anaconda. It provides the following features and several more:

- An editor includes both syntax highlighting and code completion features as well as introspection.

- View and modify environment variables from UI.

- A Help pane able to render rich text documentation for classes and functions.

Visual Studio Code

You can launch vs. code using the code command as shown in Figure 3-7.

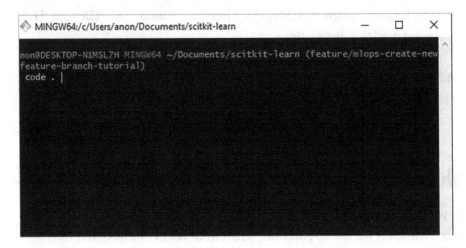

Figure 3-7. *Shortcut for launching Visual Studio Code editor from a terminal*

I'd suggest customizing the layout but at least including the Activity Bar as shown in Figure 3-8.

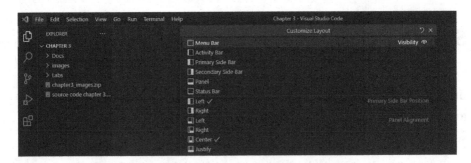

Figure 3-8. *The Activity Bar in Visual Studio Code editor*

Visual Studio Code is a source control editor from Microsoft based on the electron framework and is available for MacOS, Windows, and Linux distributions. The tool includes debugging, source control management, syntax highlighting, and intelligent code completion and operates by using extensions to add additional functionality. It is much more of a tool for large software projects and includes many extensions that allow you to interact with cloud infrastructure, databases, and services.

For example, there is an extension for Azure that allows accessing resources in the cloud. If you need to format your code, you could install one of several profile extensions or specific packages like black[2] or autopep8[3]. You search for these extensions in the activity bar and can access functionality in extensions using the keyboard shortcut **CTRL + SHIFT + P** to access the palette. We recommend at minimum you install the Microsoft Python extension or the Python Extension Package which includes linters, intellisense, and more (we'll need this when we create environments and set up tests). Figure 3-9 shows some of the Python extensions available in Visual Studio Code.

---

[2] Black is a standard code formatter used across industries. The GitHub for black is `https://github.com/psf/black`

[3] autopep8 automatically formats Python code conforming to PEP8 and can be found on GitHub at `https://github.com/hhatto/autopep8`

Figure 3-9. *Python extensions available in Visual Studio Code editor*

Cloud Notebooks and Google Colab

Cloud notebooks are a convenient way for data scientists to run code and use Python libraries in the cloud without having to install software locally. A cloud notebook such as Google Colab can be a good alternative to Visual Studio Code editor for running experiments or prototyping code. You type code into cells and can run cells in order. Figure 3-10 shows the MLOps lifecycle toolkit lab in Google Colab.

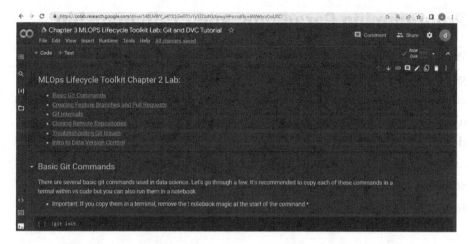

Figure 3-10. *The MLOps Lifecycle Toolkit Git and DVC Lab in Google Colab*

You can also change the theme of your notebook or connect to your GitHub through the tools ➤ settings menu. Figure 3-11 shows how to configure settings in Google Colab.

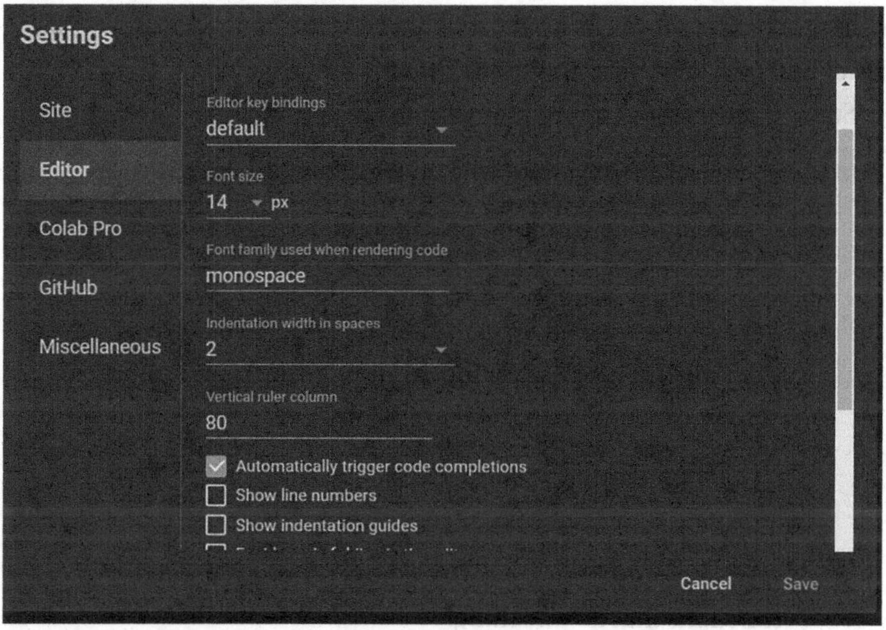

Figure 3-11. *Configuring notebook settings in Google Colab*

81

Programming Paradigms and Craftsmanship

What is craftsmanship in software? It refers to all of the high level skills you need for creating high quality data science code. Topics like naming conventions, documentation, writing tests, and avoiding common code smells all work toward writing higher quality code. Code that is high quality is often described as being "clean" which means it's more readable and maintainable, and although it may still have a higher cognitive complexity overall than other software, technical debt can be reduced by taking these topics to heart. Let's take a look at some of the elements of writing high quality data science code.

Naming Conventions and Standards in Data Science

If you don't reduce tech debt in your project, you may find yourself working overtime when something happens in production. Part of minimizing tech debt and keeping the project readable is ensuring a consistent naming convention is used for variable names, functions, class names and files, modules, and packages.

Some guidelines for naming standards are as follows:

- Use descriptive names for variables and functions.

- Consider using verbs for function names describing what your function does.

- Refer to the style guide of the language PEP8 for Python (these include advice on indentation, white space, and coding conventions).

- Use smaller line sizes for more readable code.

- Avoid long function names and functions with too many parameters – break these out into smaller functions that do one thing.

Code Smells in Data Science Code

Code smells are anti-patterns that indicate brittle code or technical debt or places in the program that could be improved. An example in Python would be using too many nested loops or hardcoding data instead of using a variable.

You might hear the term "code smell" in programming especially if your organization requires regular code reviews. During this review process, you will look for code smells. It is good practice to remove code smells when you find them as they will incur technical debt if you leave them (they may also make it more painful for other people to maintain your code when you have to hand it off to someone else or fix it yourself in the future).

A good practice is to always assume you yourself will have to maintain the code in 6 months or even a year from now and to make sure your code can be clearly understood even after you've forgotten the details of how it works.

Documentation for Data Science Teams

Most data science projects, like other software projects, are lacking in documentation. Documentation for projects can come in a number of different formats and doesn't necessarily have to mean a formal technical document; it depends on your team and the standards that have been established (if they exist). However, if coding standards don't exist, here are some recommendations for creating awesome technical documentation:

- Use doc strings without hesitation.

- Create a central repository for documentation.

- Create an acceptance criterion in tickets if you're using a board like JIRA or Azure DevOps.

- Socialize changes and ensure team members know how and where to add new documentation.

You've seen a few doc strings in the lab from the previous chapter already. We can use triple quotes underneath the function signature to describe briefly what the function does.

These doc strings are valuable because they can describe the following information:

- *What your function does*: If you find yourself trying hard to describe what your function does or realize it does more than one thing, you may want to break it up; therefore, going through this exercise of having doc strings for every function can improve quality of your code.

- *Description input, outputs, and data types*: Since languages like Python are dynamically typed, we can run into trouble by being too flexible with our data types. When you train a model on training data and forget it can take on a certain value in production, it could cause your program to crash. It's good practice to carefully consider the data types for any function arguments and, if not using type annotations, at least include the data type in the doc string.

Last but not least, make sure to update your documentation as requirements change. This leads us to the next tool in the toolkit for future proofing our code: TDD (test driven development).

Test Driven Development for Data Scientists

In data science projects especially, requirements can be fuzzy or ill-defined or missing all together in some cases. This can lead to surprises when there are huge gaps between what the user of the model or software expect and what you as the data scientist create and lead to longer release cycles or heavy refactoring down the line especially with feature engineering libraries.

One way to future-proof your code is to include tests with each function in your feature library. Of course this is time consuming, and on a real project, you may not have the time but it is strongly recommended. It only takes an hour or two to set up tests in Pytest or Hypothesis and create fixtures, and if you're using asserts already in your code, you can use these as the basis for your tests, and it will save you time if you need to debug your code in production. Figure 3-12 shows how to select a testing framework for TDD.

Figure 3-12. *Selecting a testing framework*

You may get import errors shown in Figure 3-13.

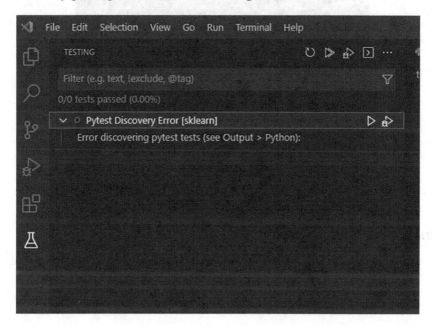

Figure 3-13. *Import errors are common when setting up Pytest in Visual Studio Code*

Once you fix the import errors, you can see tests by clicking the Testing icon in the Activity Bar and clicking run. A test that passes will have a green check mark to the left. You can run multiple tests at the same time. In the MLOps toolkit lab, you can create your own unit tests and fixtures (a way of passing data to tests) and play with this feature to incorporate testing into your own data science projects. Figure 3-14 shows how to run tests in Visual Studio Code.

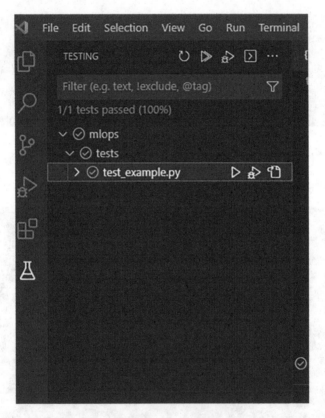

Figure 3-14. *Running tests in Visual Studio Code*

From Craftsmanship to Clean Code

There are many guidelines and principles for writing "clean code," and as you become better developers, you will come to recognize code when it is clean. In data science, clean code is often an afterthought and often only comes after translating an ad hoc analysis into something worthy for production. However, here are several principles that a data scientist can use to reduce technical debt and write cleaner, more readable code:

- Be consistent! Consistency is key especially when it comes to naming variables.

- Use separate folders for feature engineering, data engineering, models, training, and other parts of the workflow.

- Use abstraction: Wrap low level code in a function.

- If your functions are too long, break them up; they probably do more than one thing violating the *SOLID principle* of single responsibility.

- Reduce the number of parameters you use in your functions if possible (unless maybe if you're doing hyper-parameter tuning).

- Wrap lines and set a max line length in your editor.

Model Packages and Deployment

Data science software consists of a number of independent modules that work together to achieve a goal. For example, you have a training module, a feature engineering module, maybe several packages you use for missing values, or LightGBM for ranking and regression. All of these modules share something in common: You can install them, deploy them, and import them as individual deployable units called packages.

Choosing a Package Manager

Packages can consist of bundles of many modules, files, and functionality that are maintained together and are usually broader in scope than a single file, function, or module. In Python, you can use packages using a Conda or Pip or other package manager, but it's important to understand how to create your own python packages.

Setting up Packages in VS Code, use the command palette—CTRL + SHIFT + P keyboard shortcut (ensure to hold down CTRL, SHIFT, and P at the same time)—and select Python Create Environment. This is part of the Python extension package you installed earlier. Figures 3-15 through 3-18 show the detailed steps for configuring a Python environment in Visual Studio Code including selecting a package manager.

Figure 3-15. *Creating a Python environment*

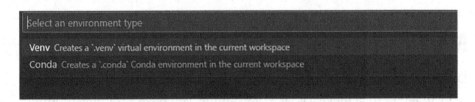

Figure 3-16. *Choosing between Conda and Virtual environment. Both are options in Visual Studio Code*

Creating environment (Show logs): Creating venv...
Source: Python (Extension)
Cancel

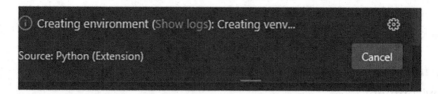

Figure 3-17. *Visual Studio Code creating a new environment*

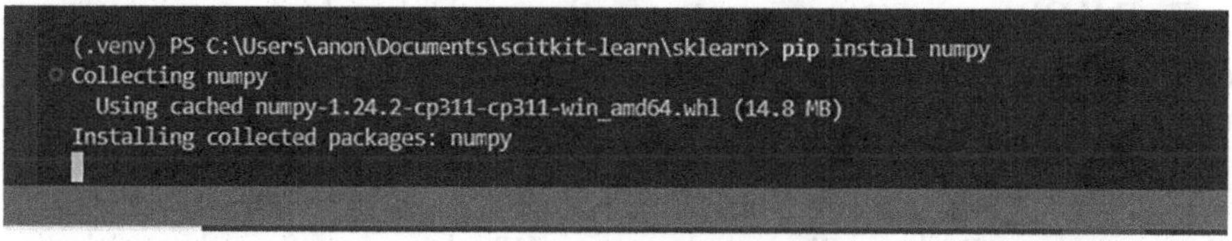

(.venv) PS C:\Users\anon\Documents\scitkit-learn\sklearn> pip install numpy
Collecting numpy
Using cached numpy-1.24.2-cp311-cp311-win_amd64.whl (14.8 MB)
Installing collected packages: numpy

Figure 3-18. *Once the environment is activated, you can install packages using your chosen package manager*

Anaconda

What is Anaconda? Well, it's not a snake. Anaconda instead is bigger than any one tool and is an ecosystem unto itself. There's a virtual environment tool called Conda which is extremely popular on data science teams. It provides several commands for package management including the following:

- conda create
- conda install
- conda update
- conda remove
- conda info
- conda search
- conda config
- conda list

The command you'll use most often to create an environment with packages is given in Listing 3-2:

Listing 3-2. Conda create command for creating a new Conda environment

```
conda create --prefix ./envs matplotlib=3.5 numpy=1.2
```

For MLOPs, we want to go a step further and take a look at some more general package managers and their features.

Pipenv: Pipenv, which we'll use in our MLOps toolkit lab, tries to bring best in breed (bundler, composer, npm, yarn, and cargo) in package management to Python. Pipenv also treats Windows as a first class citizen which makes it ideal for some business environments. You don't have to worry about low level details of creating a virtualenv for your projects as pipenv handles this for you and even auto-generates the Pipfile describing package versions and Pipfile.lock which is used for deterministic builds. Since reproducibility of experiments is an important aspect of MLOps, deterministic builds are ideal especially for large projects where you have to juggle multiple versions of packages.

An example installing the Pandas package would be given in Listing 3-3.

Listing 3-3. pipenv command for creating a new Python environment

```
pipenv install pandas
```

You will then notice Pandas has been added to the Pipfile.

Installing Python Packages Securely

Have you ever been working on a model and realized you need to install xgboost or PyTorch or some other library? It worked before but this time the computer beeps and dumps a massive error log on your screen. You spend 3 hours debugging and searching on Stackoverflow for a solution only to realize the recipe only works for Windows, not Mac!

What should you do? Use Python environments. Python environments can save you a headache by providing isolation between software dependencies. We'll show you how to set this up in the next chapter. Once you set up a Python environment, you may notice you spend less time installing and managing Python package dependencies which frees up more time to work on data science tasks.

Navigating Open Source Packages for Data Scientists

Open source software packages are released under a license (typically permissive or copyleft like GPL) that allows its users to maintain control over using and accessing the software as well as distributing, studying, and changing. Many projects you use in data science are open source such as Scikit-Learn, PyTorch, and TensorFlow and can be found on GitHub.

Technical consideration when using open source software packages in data science are the following:

- PyPi and similar repositories can contain malware, and so packages should be trusted or scanned first (see Snyk[4]).

- Open source may be maintained by a community of dedicated volunteers so patches and updates may be at whim of the maintainer.

- Copyleft and other licensing may pose challenges for building enterprise software since you need to release the software under the same license (since software is often distributed as binaries).

---

[4] You can read more about the Snyk project at https://docs.snyk.io/ manage-issues/introduction-to-snyk-projects

Common Packages for MLOps

Finally, we have enough knowledge to cover the central topic of this chapter which is packages specific to MLOps. Each of these packages provides pieces of the MLOps lifecycle such as experimentation, orchestration, training acceleration, feature engineering, or hyper-parameter tuning. We can broadly separate these packages into two camps: ModelOps and DataOps.

DataOps Packages

DataOps is a collection of best practices, processes, and technologies borrowed from Agile software engineering that are designed to improve metrics like data quality, efficient data management, and continuous data delivery for data science and more broadly analytics. We need DataOps practices and experts when we're in the data engineering part of the MLOps lifecycle. Still, there are many concepts unique to MLOps such as feature groups and model registries that typical data engineering solutions do not have. In the following, we've compiled some of the tools you might encounter when working in the first stages of the MLOps lifecycle: data collection, data cleaning, feature engineering, and feature selection.

Jupyter Notebook

Jupyter notebooks as mentioned are a useful alternative to a local code editor like Visual Studio Code. You can use notebooks for prototyping code and running experiments. However, for MLOps, a Python script is preferable to a notebook for code for a number of reasons. For example, when you source control a Jupyter notebook, it is actually a JSON file that contains a combination of source code, text, and media output. This makes it more difficult to read the raw file compared to a Python script where you can read line by line.

Python scripts are also a standard way to represent code outside of data science, and you can use many different code editors from Visual Studio Code to text-based source code editors like Sublime Text, but beyond maintaining and readability, writing code as a script enables you to create larger software projects because your code can be organized into modules, packages. This structure is very important and enables you to understand the way the project is organized, reuse code, set up tests, and use automated tools like linters that make the software development process more efficient. Therefore, I hope you will consider using Python scripts with a code editor of your choice as opposed to Jupyter notebooks for production code.

JupyterLab Server

If you do insist on using Jupyter notebooks, there are a number of environments available. One environment we already mentioned was Google Colab, but if you want to run your notebook locally and have a customizable environment that could also be deployed as a service, you might consider JupyterLab.

JupyterLab server is a Python package that sits between JupyterLab and Jupyter Server and provides RESTful APIs and utilities that can be used with JupyterLab to automate a number of tasks for data science and so is useful for MLOps. This also leads us to another widely used platform for MLOps that also comes with a notebook-based environment.

Databricks

Databricks was created by the founders of Apache Spark, an open source software project for data engineering that allows training machine learning models at scale by providing abstractions like the PySpark dataframe for distributed data manipulation.

Databricks provides notebooks, personas, SQL endpoints, feature stores, and MLFlow within its PaaS offering which is also available in multiple cloud vendors including Azure and AWS with their own flavor of Databricks.

Besides MLFlow, a vital tool for an MLOps engineer to track model metrics and training parameters as well as register models and compare experiments, Databricks has a concept of a delta lakehouse where you can store data in parquet format with a delta log that supports features like time travel and partitioning.

We'll mention this briefly, but it could have its own chapter since this is a massive topic. Koalas is a drop-in solution although not 100% backward compatible with Pandas (of course, there's a lag between when a feature is supported in Pandas and when it becomes generally available in Pandas for Spark), but this is a great tool to add to your toolkit when you need to scale your workflow. While doing development in PySpark, you don't have to re-write all of your code; you use following import at the top of your file and use it like you would Pandas.

Dask: Dask is another drop-in solution for data wrangling similar to Pandas except with better support for multiprocessing and large data sets. The API is very similar to Pandas, but unlike Koalas or Pandas API for Spark, it is not really a drop-in solution

Modin: While Dask is a library that supports distributed computation, Modin supports scaling Pandas. It supports various backends including ray and Dask. Again, it's not 100% backward compatible and has a much smaller community than Pandas, so use with caution on a real project.

ModelOps Packages

ModelOps is defined by Gartner as " focused primarily on the governance and lifecycle management of a wide range of operationalized artificial intelligence and decision models, including machine learning, knowledge graphs, rules, optimization, linguistic, and agent-based models." Managing models is difficult in part because there's code and data and many different types of models as we've seen from reinforcement learning to deep learning to shallow models in scikit-learn and bespoke statistical models.

We list some of the most popular tools for ModelOps in the following that you may encounter when you work in the later half of the MLOps lifecycle which includes model training, hyper-parameter tuning, model selection, model deployment, model management, and monitoring.

Ray[5]*:* Ray is a great tool for reinforcement learning; it is based on the actor model of distributed computation, in computer science,[6] and allows you to use decorators to scale out functions which is convenient when you don't want to rewrite a lot of code.

KubeFlow[7]*:* KubeFlow is another open source machine learning tool for end to end workflows. It is built on top of Kubernetes and provides cloud-native interfaces for building pipelines and containerizing various steps of the machine learning lifecycle from training to deployment.

Seldon[8]: Have you ever been asked to deploy your machine learning models to production? First of all, what does that even mean? There are many ways to deploy a model. You could put it in a model registry, and you could containerize your model and deploy it to Ducker Hub or another container registry, but for some use cases especially if an end user is going to be interacting with your model on demand, you'll be asked to expose the model as an API.

Building an API is not a trivial task. You need to understand gRPC or REST and at least be familiar with a framework like Flask if you're using Python. Fortunately, there are tools like Seldon that allow you to shortcut

---

[5] The Ray framework documentation can be found at `https://docs.ray.io/en/latest/`

[6] Agha, G. (1986). *Actors: A Model of Concurrent Computation in Distributed Systems.* `https://apps.dtic.mil/sti/pdfs/ADA157917.pdf`

[7] The KubeFlow project documentation can be found at `www.kubeflow.org/docs/`

[8] The Sledon project documentation can be found at `https://docs.seldon.io/projects/seldon-core/en/latest/index.html`

some of these steps and deploy models as gRPC or REST endpoints. Seldon in particular offers two models for servers: reusable and nonreusable. The definition of each is stated in the following.

- *Reusable model servers:* These are prepackaged model servers. You can deploy a family of models that are similar to each other, reusing the server. You can host models in an S3 bucket or blob storage account.

- *Nonreusable model servers:* This option doesn't require a central model repository, but you need to build a new image for each model as it's meant to serve a single model.

This leads us to the standard solution right now for registering your model, MLFlow. You had to create your own model storage and versioning system and way to log metrics and keep track of experiments. All of these important model management tasks (ModelOps) are made easier with MLFlow.

Model Tracking and Monitoring

MLFlow[9] is the standard when it comes to creating your own experimentation framework. If you've ever developed loss plots and kept track of model metrics and parameters during hyper-parameter tuning, then you need to incorporate MLFlow into your project.

You can set up the MLFlow infrastructure as a stand-alone or part of Databricks (the original developers). We'll see this in action in later chapters.

---

[9] MLFlow project documentation can be found at `https://mlflow.org/docs/latest/index.html`

HyperOpt[10]: Hyperopt is a framework for Bayesian hyper-parameter tuning, often done after the cross validation step but before training a model on the entire data set. There are also many algorithms available depending on the type of parameter search you need to do including the following:

- Random search

- Tree of Parzen Estimators

- Annealing

- Tree

- Gaussian Process Tree

Horovod[11]: Horovod is a distributed deep learning framework for TensorFlow, Keras, PyTorch, and Apache's MXNet. When you need to accelerate the time it takes to train a model, you have the choice between GPU accelerated training and distributed training. Horovod is also available on Databricks and can be a valuable tool for machine learning at scale.

Packages for Data Visualization and Reporting

If you've ever had to do a rapid EDA or exploratory data analysis, you know how tedious it can be to have to write code for visualizations. Some people like writing algorithms and don't like visualization, whereas others who are good at libraries like Matplotlib or Seaborn become the de facto visualization experts on the team.

---

[10] The Hyperopt project can be found on GitHub at `https://github.com/hyperopt/hyperopt`

[11] The Horovod project source code can be found at `https://github.com/horovod/horovod`

From an MLOps perspective, visualizations can be an "odd one out" in a code base and are difficult to deploy since creating interactive plots and dashboards requires special knowledge and tools. You should at least be familiar with a couple tools beyond Matplotlib for exploratory data analysis including the following:

- *Dash*[12]: Python library for creating interactive dashboards

- *PowerBI:* Visualization software from Microsoft. Useful for data science since you can embed Python and deploy to cloud

Lab: Developing an MLOps Toolkit Accelerator in CookieCutter

This lab is available on the Apress GitHub repository associated with this book. You will see in Chapter 3 the mlops_toolkit folder. We will use a package called cookiecutter to automate the process of setting up tests, train, data, models, and other folders needed in future chapters. Figure 3-19 shows the toolkit folders.

---

[12] The Dash project can be found on GitHub at https://github.com/plotly/dash

Figure 3-19. *MLOps toolkit folder structure*

You might be wondering what the point of having a template like this is. The primary reason is it goes toward establishing standards and code structure that borrows from experience across several industries. This pattern is tried and proven, and although it is slightly opinionated on use of testing framework and names of folders, you can easily customize it to your purposes.

We'll do exactly this by installing several packages that can support other stages of the MLOps lifecycle such as model training, validation, hyper-parameter tuning, and model deployment. The steps for setting up the lab are as follows:

Step 1. Clone the project locally and run the following command to open vs code:

Listing 3-4. Shortcut for opening Visual Studio Code[13]

```
code .
```

Step 2. Start a new vs. code terminal session (here we're using PowerShell but you can also use Bash) and cd into the mlops_toolkit directory. Figure 3-20 shows the root directory.

Figure 3-20. *Root directory for MLOps toolkit supplementary material*

Step 3. Clear the screen with the clear command and type as shown in Listing 3-5.

Listing 3-5. Installing Pandas package with a specific version number using Pipenv

```
pipenv install pandas~=1.3
```

Step 4. Check the Pipfile containing the following lines.

Step 5. Repeat steps 2–3 for the following packages: numpy, pytest, hypothesis, sckit-learn, pyspark, and mlflow. By default, the latest versions will be installed, but we recommend using the ~ operator with a major. minor version to allow security patches to come through. The output is shown in Figure 3-21.

---

[13] Tips and Tricks for Visual Studio Code https://code.visualstudio.com/docs/getstarted/tips-and-tricks

```
PS C:\Users\anon\mlops_lifecycle\fake_corp_toolkit> pipenv install mlflow~=2.2
Courtesy Notice: Pipenv found itself running within a virtual environment, so it will automatically use that environment, instead of creat
ing its own for any project. You can set PIPENV_IGNORE_VIRTUALENVS=1 to force pipenv to ignore that environment and create its own instead
. You can set PIPENV_VERBOSITY=-1 to suppress this warning.
Creating a Pipfile for this project...
Installing mlflow~=2.2...
Resolving mlflow~=2.2...
Installing...
[==  ] Installing mlflow...
```

Figure 3-21. *The result of installing some Python packages with pipenv*

Step 6. ***CTRL + SHIFT + P*** to open the vs code command palette. Type python and choose pytest in the dropdown and select/tests folder.

Step 7. Click the tests icon in the Activity Bar and run all tests by clicking the "run" button.

Step 8. Run the following command with the custom name of your project.

Step 9. Cd into the folder you created and customize it to your own data science project. Here I used main_orchestrator.py for the file name.

Step 10. Python main_orchestrator.py should print a message to the screen as shown in Figure 3-22.

```
PROBLEMS    OUTPUT    TERMINAL                                                                          powershell - fake_corp_toolkit
> ∨ TERMINAL
    PS C:\Users\anon\mlops_lifecycle\fake_corp_toolkit> python .\main_orchestrator.py
    Hello, I'm in the main orchestrator of the project. MLOps is fun!
    PS C:\Users\anon\mlops_lifecycle\fake_corp_toolkit>
```

Figure 3-22. *Running the main orchestrator should print a message to your screen*

Step 11. Go through the Git fundamentals lab again if necessary, and add code and data version control by running two commands in a terminal (works both in PowerShell and Bash) as given in Listing 3-6:

Listing 3-6. Initializing source and data version control commands in a repo[14]

```
git init
dvc init
```

That's it! Not so bad and we've already set up tests, our very own custom monorepo, installed packages to support various stages of the lifecycle, and know how to set up code version control and data version control. In the next chapters, we'll go through the gritty details of MLOps infrastructure, model training, model inference, and model deployment, developing our toolkit further.

Summary

In this chapter, we gave an introduction to several tools for MLOps and data science including version control both for source code and data. We also talked about the differences between Jupyter notebooks and Python scripts and why Python scripts are the preferred format for MLOps. We looked at code editors like Visual Studio Code for working with Python scripts and talked about some of the tools, packages, and frameworks you may encounter in an MLOps workflow. Here is a summary of what we learned:

- Data and Code Version Control Systems

- Model Development and Training

- Model Packages and Deployment

- Model Tracking and Monitoring

In the next chapter, we will shift our attention to infrastructure and look at how we can begin to use some of the tools discussed in this chapter to build services to support the various stages of the MLOps lifecycle.

---

[14] DVC User Guide: `https://dvc.org/doc/user-guide`

CHAPTER 4

Infrastructure for MLOps

This chapter is about infrastructure. You might think of buildings and roads when you hear the word infrastructure, but in MLOps, infrastructure refers to the most fundamental services we need to build more complex systems like training, inference, and model deployment pipelines. For example, we need a way to create data stores that can store features for model training and servers with compute and memory resources for hosting training pipelines. In the next section, we will look at a way we can simplify the process of creating infrastructure by using containers to package up software that can easily be maintained, deployed, and reproduced.

Containerization for Data Scientists

Containers have had a profound impact on the way data scientists code; in particular, it makes it possible to quickly and easily spin up infrastructure or run code inside a container that has all of the software, runtimes, tools, and packages you need to do data science bundled inside.

Why is this a big deal? As you've probably experienced in the previous chapter where we used Python environments to isolate packages and dependencies, a lot of problems with configuring and managing multiple packages become manageable with containerization. With simple

D. Sorvisto, *MLOps Lifecycle Toolkit*, https://doi.org/10.1007/978-1-4842-9642-4_4

103

environments like Conda, you could manage multiple versions and with package managers like Pipenv, you had access to a Pipfile which contained all of the configuration you needed to manage your environment.

Now imagine you need more than just Python; you might have different runtime requirements. For example, maybe parts of your data science workflow require R packages and so you need the R runtime. Maybe you also have to manage multiple binaries from CRAN and have your code "talk" to a database which itself has dependencies like a JVM (Java virtual machine) or specific configuration that needs to be set.

Unless you have a strong background in IT, managing all of these configurations, runtimes, toolchains, compilers, and other supporting software becomes tedious and takes away from time you can spend on data science.

There's another problem: portability. Imagine you have a package that requires Mac but you're on Windows. Do you install an entire OS just to run that software? Containers solve this problem by allowing you to build once and run anywhere. They also make it pull new containers and swap out components of your machine learning system with ease. Let's take a deep dive into one of the most popular container technologies: Docker.

Introduction to Docker

Docker is a platform as a service that uses OS-level virtualization to encapsulate software as packages called containers. The software that hosts the containers is called the Docker Engine and is available for a number of platforms including Linux, Windows, and MacOS with both free and paid licensing. Figure 4-1 shows how containers run on top of the Docker Engine using OS-level virtualization.

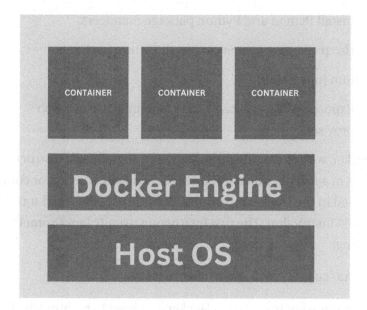

Figure 4-1. *How containers run using OS-level virtualization*

Anatomy of the Docker File

Okay, so we know what Docker is but how do we use it? Let's say you want to create our own Jupyter based data science lab. You've probably installed Jupyter before, but could you write down a recipe that is reproducible? You might start by noting what operating system (OS) you're using, installing dependencies like Python and pip, and then using pip to install Jupyter lab. If you've read the Jupyter lab documentations, then you probably also know you need to expose some default ports so you can launch and access your notebook from a web browser. If you wanted to do a deep learning workflow using GPU, you might consider installing NVIDIA drivers as well.

This is a lot of work but we can write it as a series of steps:

- From your host OS, install specific software packages.

- Install drivers and low level package managers.

- Install Python and Python package managers.

- Use package managers to install Python packages.

- Run Jupyter lab.

- Expose ports so we can access Notebooks in our web browser.

In Docker, we can encode these steps as a sequence of instructions or commands in a text file called a *Docker File*. Each instruction or command gets executed in the Docker environment in the order it's read starting from the first instruction. The first instruction usually looks something like the following:

```
FROM nvidia/cuda:12.0.1-base-ubuntu20.04
```

This creates what is known in Docker as a *layer* containing the Ubuntu OS with NVIDIA's cuda drivers in case we need GPU support (if you only have a CPU on your laptop, you can still build this docker container).

Other *layers* get installed on top of this layer. In our example of installing a deep learning library, we would need to install Cuda and Nvidia drivers to have GPU accelerated training (covered in the next section). Fortunately, in Ubuntu, these are available in the form of Ubuntu packages. Next, we might want to create a dedicated working dir for all of our notebooks. Docker comes with a WORKDIR instruction. We'll call our directory /lab/

```
WORKDIR /lab/
```

We need to install data science specific Python packages for our lab environment and most important the Jupyter lab packages. We can combine this step into a single command with the RUN instruction.

```
RUN pip install \
    numpy \
    pandas \
```

```
tensorflow \
torch \
Jupyterlab
```

Finally we'll need to launch our Jupyter server and expose port 8080 so we can access our notebook in a browser. It's good practice to change the default port, but ensure it's not one that is reserved by the operating system. These steps can be accomplished using the CMD and EXPOSE instructions:

```
CMD ["jupyter", "lab", "--ip=0.0.0.0", "--port=8080",
"--allow-root", "--no-browser"]
EXPOSE 8080
```

In the next section, we will apply this theoretical knowledge of Docker by packaging all of these steps into a Docker file in the next lab and build the image. Once we build the image (a binary file) we can then run the image, creating a *container*. This distinction between an image and container might be confusing if it's the first time you've encountered the terms, but you should understand the difference before proceeding to the lab. We summarize the difference in the following since it is very important for understanding containers.

Docker file: A docker file is a *blueprint* for a Docker image; it contains a series of instructions in plain text describing how a docker image should be built. You need to first build the docker file.

Docker image: A docker image is a binary file that can be stored in the cloud or on disk. It is a lightweight, self-contained (hence the name container), executable piece of software that includes everything needed to run an application including the application code, application runtime, interpreters, compilers, system tools, package managers, and libraries.

Docker container: A docker container is a live piece of software; it's a *runtime instance* of a Docker image created when you run the image through the Docker run command.

Now that we've clarified the difference between a Docker image and a Docker container, we're ready to start building some containers. In this lab, you'll go through the previous steps in detail to create your own data science lab environment, an important addition to any MLOps engineer's toolkit.

Lab 1: Building a Docker Data Science Lab for MLOps

Step 1. We first need to install the Docker engine. Proceed to Install Docker Desktop on Windows and select your platform. We'll be using Windows 10. Download the Docker Desktop Installer for Windows. We recommend the latest version but we'll use 4.17.1.

Step 2. Right-click the Docker Desktop Installer run as admin. Ensure to check the install WSL and Desktop shortcut options in the first menu, and click next. Figure 4-2 shows Docker Desktop for Windows.

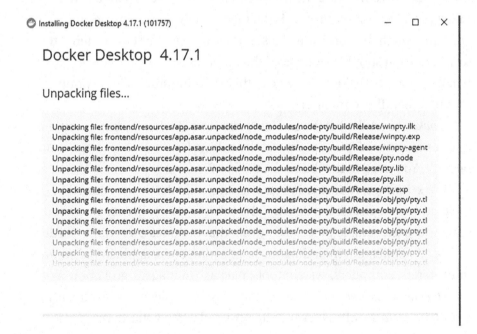

Figure 4-2. *Docker Desktop for Windows*

Step 3. Launch Docker Desktop from the Desktop icon and accept the service level agreement. Figure 4-3 shows the Docker license agreement.

Docker Subscription Service Agreement

By selecting **accept**, you agree to the Subscription Service Agreement, the Docker Data Processing Agreement, and the Data Privacy Policy.

Note: Docker Desktop is free for small businesses (fewer than 250 employees AND less than $10 million in annual revenue), personal use, education, and non-commercial open source projects. Otherwise, it requires a paid subscription for professional use. Paid subscriptions are also required for government entities. Read the FAQ to learn more.

View Full Terms [ ]   Accept   Close

Figure 4-3. *Docker license agreement*

Step 4. Use the Git clone command to clone the repo provided along with the supplementary resources. Start a new terminal session in vs code and cd into Chapter 4 Labs where you will find a file called Dockerfile (this is where you'll find the sequence of plain text instructions or recipe for building your data science lab environment).

Step 5. Run docker build -t data_science_lab . inside the directory with Dockerfile. The period at the end is important; it's called the *Docker context*.

Step 6. Build your image. Assign it the name jupyter_lab with the -t option and run the container. You can also pass in a token (we used mlops_toolkit) which will be your password for authenticating with Jupyter notebook.

```
docker build -t jupyter_lab .
docker run --rm -it -p 8080:8080 -e JUPYTER_TOKEN=mlops_toolkit
jupyter_lab
```

Did you notice anything? You should see the following splash screen. Figure 4-4 shows a general view of what you can expect to see, but note that your splash screen may look slightly different especially if you are not using Powershell.

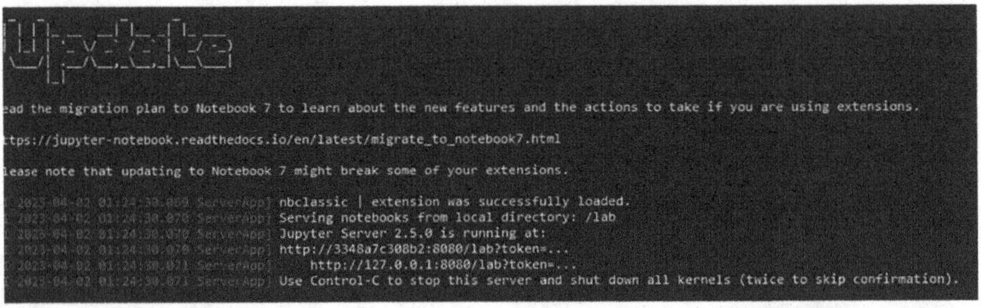

Figure 4-4. Splash screen for Jupyter Lab

Note   The PyTorch and TensorFlow wheels are around 620 and 586 MB, respectively, at the time of writing, so these are pretty large. Sometimes this can be a problem if disk space is limited. Although we won't cover it in this lab, optimizing the size of a Docker image is an interesting problem and an area of specialization within MLOps especially when working with deep learning frameworks.

Step 7. Navigate to localhost:8080/lab in a browser (note we exposed port 8080 in the Dockerfile; this is where the number comes from). Enter your token ("mlops_toolkit") and you should be redirected to the lab environment pictured in Figure 4-5.

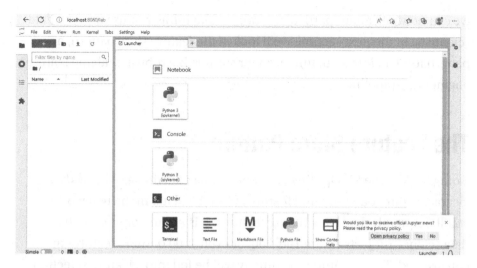

Figure 4-5. *The Jupyter lab environment*

Finally, click Python Kernel in your Lab environment to launch a Jupyter notebook. We'll use this environment in subsequent labs if you need a Jupyter notebook environment. Optionally you can also use Google Colab (Figure 4-6).

Figure 4-6. *Examples of cells in a Jupyter notebook*

You can now run notebooks inside docker and you have a reproducible data science lab environment you can use and share. We'll talk about how you can deploy this environment to the cloud in future chapters enabling you to have a shared lab environment and collaborate on projects. This is an amazing first step toward mastering data science infrastructure, and we can now talk about particular kinds of data infrastructure used by data scientists. In the next section, we'll look at the feature store pattern, a pattern for data infrastructure used for supporting robust, scalable feature engineering pipelines.

The Feature Store Pattern

Going back to the MLOps lifecycle, after data collection and basic data cleansing, our goal is to build features from this data. In the real world, you'll frequently deal with 100s of features. It is not uncommon to have data science projects where 100, 200, or even 1000 or more features are constructed. These features eventually will be fed into a feature selection algorithm, for example, when we have a prediction problem using a supervised data, we can reduce these hundreds of features to a reasonable number in many ways, for example, using Lasso or a bagging algorithms like random forest to rank features by importance for our particular problem, filtering out the ones that have little predictive value.

The feature selection process, unlike most other parts of the machine learning lifecycle, may not be automated. One reason for this is feature selection is dependent on what you're trying to model, and there may be specific features like demographic data and PII that need to be excluded even if those features have predictive value.

Although feature selection can reduce the number of features used in a model, we still need to maintain the complete gambit of features for future problems. Additionally, model accuracy deteriorates over time, the business definitions can change, and you may have to periodically rerun feature selection as you add new data sources.

So how do we store all of these features: manage different versions of features to support feature selection, hyper-parameter tuning, model retraining, and future modeling tasks that might make use of these hundreds of features at different points in the lifecycle? This is where the concept of a feature store comes into play.

Feature store: A feature store is a design pattern in MLOps that is used to centralize the storage, processing, and access to features. Features in a feature store are organized into logical groups called *feature groups*, ensuring the features are reusable and experiments are reproducible.

Implementing Feature Stores: Online vs. Offline Feature Stores

A feature store and feature groups may be implemented using a variety of data infrastructure. A model is trained on features which typically involve joining multiple normalized, disparate data sources together. These joins are expensive and, sometimes since data is not well-defined, may involve semi-joints, range joins, or window analytic functions in the mix. These queries, which are executed on remote data store infrastructure, need to support both low latency queries at prediction time and high throughput queries on years of historical data at training time.

To make matters more complex, features may not be available at prediction time or may need to be computed on the fly, possibly using the same code as in the training pipeline. How do we keep these two processes in sync and have data infrastructure support both online and offline workflows requiring low latency and high throughput?

This is a hard problem in MLOps but understanding the types of *data infrastructure* used to implement a feature store. Let's look at some of this data infrastructure we can use for implementing feature stores.

Lab: Exploring Data Infrastructure with Feast

Feast is an open source project (Apache License 2.0 free for commercial use) that can be used to quickly set up a feature store. Feast supports both model training and online inference and allows you to decouple data from

ML infrastructure, ensuring you can move from model training to model serving without rewriting code. Feast also supports both online and offline feature stores.

You can also build your own feature store using docker and installing the underlying database services yourself. At the end of this lab, there is also an exercise so you aren't just following instructions, but first run the following commands:

Step 1. Open code and start a new terminal session in PowerShell or Bash. Install Feast:

```
pipenv install feast
pipenv shell
```

Step 2. Run the following command to initialize the Feast project. We will call our feature store pystore:

```
feast init pystore
cd pystore/feature_repo
```

Step 3. Look at the files created:

data/ are parquet files used for training pipeline.

example_repo.py contains demo feature definitions.

feature_store.yaml contains data source configuration.

test_workflow.py showcases how to run all key Feast commands, including defining, retrieving, and pushing features.

Step 4. You can run this with python test_workflow.py. Note on Windows we had to convert our paths to raw strings to get this to work (see the code for Chapter 4). Figure 4-7 shows the result of running the test script.

```
Created entity driver
Created feature view driver_hourly_stats
Created feature view driver_hourly_stats_fresh
Created on demand feature view transformed_conv_rate
Created on demand feature view transformed_conv_rate_fresh
Created feature service driver_activity_v2
Created feature service driver_activity_v1
Created feature service driver_activity_v3
```

Figure 4-7. *Running the test script locally*

Step 5. Run Feast apply (inside pystore directory); this will register
entities with Feast. Figure 4-8 shows the result of running this command.

```
Updated feature view sensor
        stream_source:  -> name: "sensor_push_source"
type: PUSH_SOURCE
data_source_class_type: "feast.data_source.PushSource"
batch_source {
  name: ".\\data\\sensor.parquet"
  type: BATCH_FILE
  timestamp_field: "timestamp"
  file_options {
    uri: ".\\data\\sensor.parquet"
  }
}
```

Figure 4-8. *Running Feast apply command*

Note, you should see two components with temperature and pressure
measurements generated in your final feature store pictured in the
following. That's it! You've created your first feature store for an IoT data
set. Figure 4-9 shows the expected output.

	component_id	pressure	temperature
0	1	1545.093262	1973.979370
1	2	1795.720947	1719.880371

Figure 4-9. *Expected output for pressure and temperature readings*

115

Now as promised, here is an exercise you can do to get a feel for a real MLOps workflow.

Exercise

Being able to iterate and make changes to feature definitions is a part of MLOps since features rarely stay static. In a production environment, these types of anomalies should be caught automatically.

Exercise 1. Modify the notebook and rerun the lab to fix the pressure and temperature features so that they're in a more reasonable range for pressure (measured in Kilopascals) and temperature (measured on the Kelvin scale).

Hint   You may need to do some research on what the right range looks like and figure out where in the code you should make the change.

Dive into Parquet Format

You also may have noticed the format we are storing our data. We used the parquet extension as opposed to the more common csv which you're probably already familiar with. So what is the difference between a parquet and a csv and why might we prefer to store files in parquet format at all?

The difference is in the size and efficiency of the format. While parquet is highly efficient at data compression (it is a binary file) meaning the file sizes are much smaller, unlike csv, parquet format encodes the data and schema for fast storage and retrieval. You might also use Parquet format with libraries like Apache Arrow which can make reading a large csv file several times faster. There is also a difference in how the data is stored.

In parquet format, data is stored in a columnar format, whereas csv is row oriented. For data science code, columnar data store is preferred since only a small subset of columns are used for filtering, grouping, or aggregating the data.

Although knowledge of every possible data format isn't required, you should be aware as an MLOps engineer that you can optimize your code for speed and efficiency simply by changing the format to one that better matches your workflow. In the next section, we'll take a look at another way to optimize for speed: hardware accelerated training.

We just took a deep dive into containers and data infrastructure, but if you're a pure data scientist without a background in IT, then you might be wondering do I really need to know how to work with low level data infrastructure and become an expert in containers to do MLOps for my own projects?

The answer depends on the use case, but in general, there are cloud services available for each stage of the MLOps lifecycle. For example, you can use Databricks if you want an end-to-end machine learning platform and add components as needed by integrating with other cloud services, for example, PowerBI, if you need a reporting solution, Azure DevOps if you need to build CI/CD pipelines to deploy your code, and maybe even an external data storage like AWS or Azure data lake to store your models, artifacts, and training data sets. You technically should know about parquet, but in this example, you could use Delta table format which in uses Parquet under the hood for storing data but also gives you a delta log and APIs for working with this format, so the low level details are abstracted for you, leaving more time for data science. In the next section, we'll take a deeper dive into some of the cloud services available while trying to remain agnostic about specific platforms like AWS, Azure, and Google Cloud.

Hardware Accelerated Training

Many times in data science, we are dealing with big data sets. Training sets can total gigabytes, terabytes, and with the rise of IoT data even petabytes of data. In addition to big data, many workflows, especially ones requiring deep learning like transfer learning, can be extremely intensive and require GPU accelerated training.

Training or even fine-tuning a large language model like BERT on commodity hardware using only a CPU can take days. Even if you're not training a large language model from scratch, some model architectures like recurrent neural networks take a long time to train. How do we accelerate this training? We have two options: distributed training and GPU accelerated training. First, let's discuss some of the major cloud service providers before jumping into distributed training.

Cloud Service Providers

There are several major cloud service providers. The big 3 are Azure, Amazon Web Services, and Google Cloud. Each of the three has machine learning service offerings and provides compute, networking, and data storage services. For example, Amazon Web Services has s3 buckets and Azure has blob storage. For end-to-end machine learning, Amazon Web Services offer SageMaker, while Azure has Azure Machine Learning service. There are other services for end-to-end machine learning as well and distributed training like Databricks which is offered in all three of the cloud service providers. There are differences between the different services, for example, Databricks integrates with MLFlow, whereas SageMaker has its own Model registry, but there is a difference in the platform: not the cloud service provider. You can also deploy your own containers in all three cloud service providers. For example, if you want to deploy your own Airflow instance to Kubernetes, all three offer their own version of Kubernetes with differences in cost for compute, storage, and tooling. In the next section, we'll take a look at distributed computing in some of these cloud service providers.

Distributed Training

All of the code we've run so far has been executed on a single machine. If you're using a laptop or workstation, you can interact with the physical hardware, but if you're running inside a cloud environment like Google Cloud, Azure, AWS (Amazon Web Services), Google Colab, or Databricks, the hardware infrastructure on the backend may not be so obvious and may actually be hidden from you. For example, in Databricks, you can configure a cluster, a collection of worker nodes and driver nodes which are themselves individual virtual machines complete with their own CPU or GPU for compute and a certain configurable amount of working memory and disk space.

The advantage of using multiple VMs when training is straightforward: More VMs mean more CPU or GPUs available which means model training can be accelerated. If you've ever written Pandas code that attempted to read in a large csv file and experienced out of memory errors, then you've probably already thought about increasing the memory available through out of core (spilling to disk) like Dask, but another option is to run your code on a distributed environment like Databricks.

You can take a look at the supplementary code provided with this chapter for an example of configuring Horovod for distributed training.

You can make a free account on Databricks community edition to try out Databricks, but we recommend you use an Azure cloud subscription for full functionality. The steps to get a Databricks account (which you can later convert to a full featured account) are as follows:

1. In a browser, navigate to `https://community.cloud.databricks.com/login.html`.

2. Click sign up and create a free account. Figure 4-10 shows how to register a Databricks account.

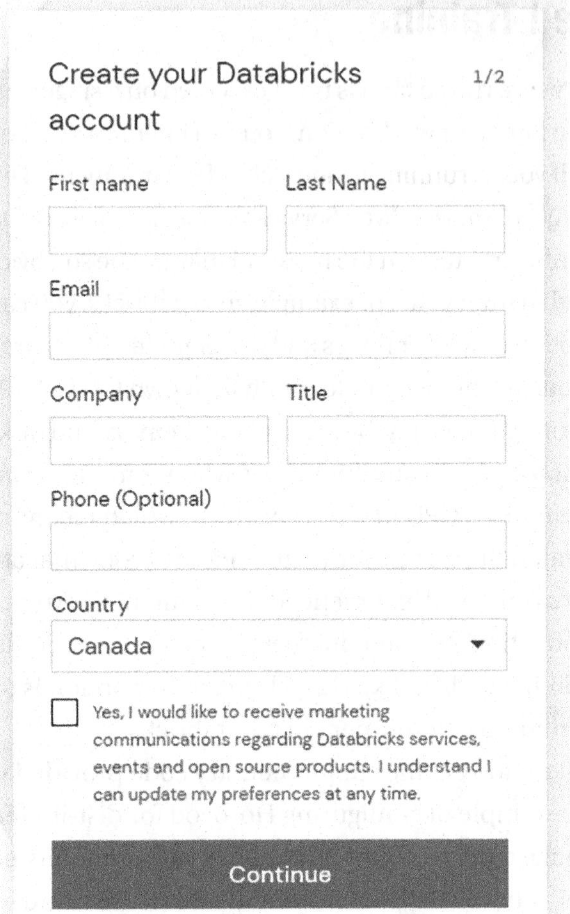

Figure 4-10. *Registering a Databricks account*

Click continue and make sure to select community edition at the bottom; otherwise, choose a cloud provider (AWS, Azure, or Google Cloud as shown in Figure 4-11):

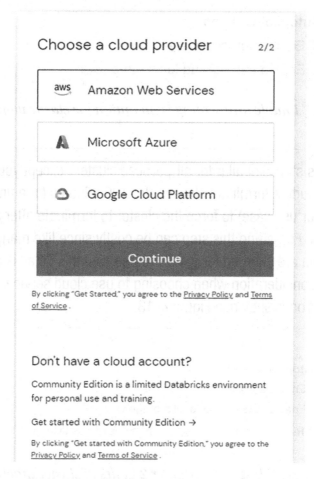

Figure 4-11. *Choosing a cloud provider*

3. In the workspace, create a job cluster. Databricks distinguishes between two types of clusters: all purpose (interactive) and job clusters.

4. Click the cluster creation and edit page; select the Enable autoscaling checkbox in the Autopilot Options box (Figure 4-12).

Autopilot Options

☑ Enable autoscaling ❷

☐ Enable autoscaling local storage ❷

Figure 4-12. *Enable autoscaling is an option for elastic workflows*

Note   This step is similar for all-purpose clusters except you will want to include a terminate clause after 120 minutes (or a timeout that fits your use case) to force the cluster to terminate after a period of inactivity. Forgetting this step can be costly since like many cloud services you are charged for what you use, and this detail is an important consideration when choosing to use cloud services. The timeout option is shown in Figure 4-13.

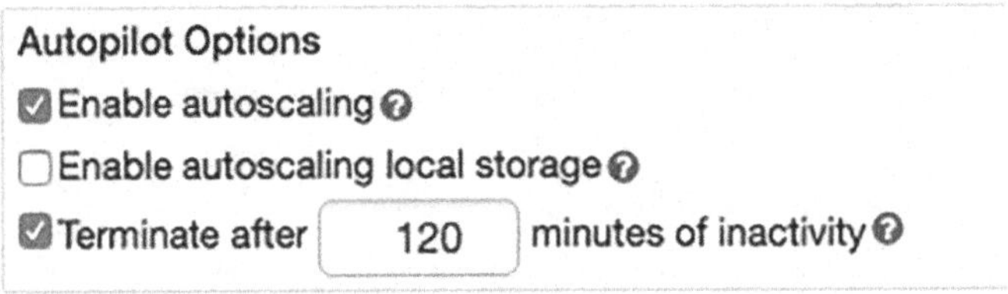

Autopilot Options

☑ Enable autoscaling ❷

☐ Enable autoscaling local storage ❷

☑ Terminate after [120   ] minutes of inactivity ❷

Figure 4-13. *Enabling timeout after 2 hours of cluster inactivity*

To attach a cluster to a notebook in Databricks, follow these steps:

1. Create a new notebook in your workspace.

2. Click the "Connect" button in the top-right corner of the notebook (Figure 4-14).

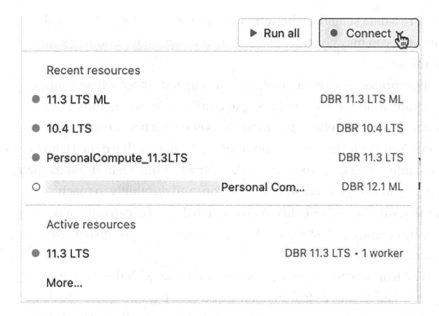

Figure 4-14. Connect button to attach a notebook to a cluster

Select the cluster you just created in the previous step.

Once the cluster is attached, you are able to run your code on the cluster, taking advantage of the many workers available for distributed workflows. You can configure the number of workers in your cluster and enable autoscaling for larger workflows. The notebook will connect to the cluster automatically. You can also detach the cluster from the notebook by clicking the "Detach" button in the top-right corner of the notebook. You can optionally copy paste code provided in the next section if you want to try this out.

Optional Lab: PaaS Feature Stores in the Cloud Using Databricks

You may have noticed when using Feast there were a lot of steps and you had to dive deep into the gritty details of data engineering infrastructure and even understand different types of data formats like parquet vs. csv

123

If you're a data scientist who wants some of those details abstracted from you, you may consider a Platform as a Service for building your feature store.

Databricks provides a machine learning workspace where feature stores are available without having to configure infrastructure. These feature stores use delta tables in the backend which rely on the open source Parquet format, a column oriented format for big data. Delta tables also come with a delta log that can keep track of transactions on the data, bringing atomicity, consistency, isolation, and durability to machine learning workflows (so-called ACID properties). You can build a feature store by creating a cluster with the ML runtime[1] (12.1 is the latest at the time of writing).

The feature store client allows you to interact with the feature store, register data frames as feature tables, and create training sets consisting of labeled data and training data for use in training pipelines. Databricks also has online feature stores for low latency inference pipelines.

Scaling Pandas Code with a Single Line

If you use Pandas regularly for data wrangling tasks, you may have encountered memory errors. Typically dataframes blow up in memory up at least 2x and sometimes more compared to their size on disk which means if you have a very large csv file, reading that csv file may trigger some out of memory errors if your workflow relies on Pandas. Fortunately, the Pandas on Spark library (formerly Koalas) allows you to write Pandas code to create Spark dataframes and register them in the feature store without having to learn the Spark API. You can import this library in Databricks with the following line (called a drop-in solution).

```
from pyspark import pandas as ps
```

---

[1] Databricks ML runtime documentation can be found at https://docs.databricks.com/runtime/mlruntime.html

We've provided an option notebook lab for you called Chapter 4 Lab: Scaling Pandas Workflows provided with this chapter. You can import your notebook into your Databricks workspace and execute the code to get hands-on experience with scaling Pandas code.

Since Databricks requires a cloud subscription, you don't need to complete this lab to understand the rest of the material in this chapter or the rest of the book; however, many organizations use Databricks for MLOps, and knowledge of PySpark, the Pandas on Spark library, clusters, and notebooks may be valuable in an MLOps role. You can import a notebook by clicking Workspace or a user folder and selecting Import as pictured (Figure 4-15):

Figure 4-15. *Importing a notebook in your workspace*

GPU Accelerated Training

GPU accelerated training means using a GPU to help reduce the runtime in training deep learning algorithms. While CPUs are latency optimized, GPUs are bandwidth optimized. Since deep learning involves tensor operations and matrix multiplications which can be implemented on a GPU, these operations can be sped up by using a framework that is GPU aware both because of the data parallelism and the higher bandwidth afforded by a GPU.

One exciting change to the TensorFlow package is that since version 2.1, TensorFlow and TensorFlow-gpu have merged. If you require a version of the TensorFlow package with version <= 2.1, then you can install TensorFlow-gpu as per the following otherwise you may substitute TensorFlow-gpu with TensorFlow.

In your Jupyter lab environment, you can make your notebook GPU aware by using TensorFlow's TensorFlow-gpu package (other deep learning frameworks such as PyTorch require code changes). The steps for configuring GPU awareness in TensorFlow are listed in the following:

1. Uninstall your old TensorFlow.

2. Edit your Dockerfile and add the TensorFlow package to the RUN pip install command (note if for backward compatibility, you require TensorFlow < 2.1, and then use the older TensorFlow-gpu package instead). Figure 4-16 shows the informational message.

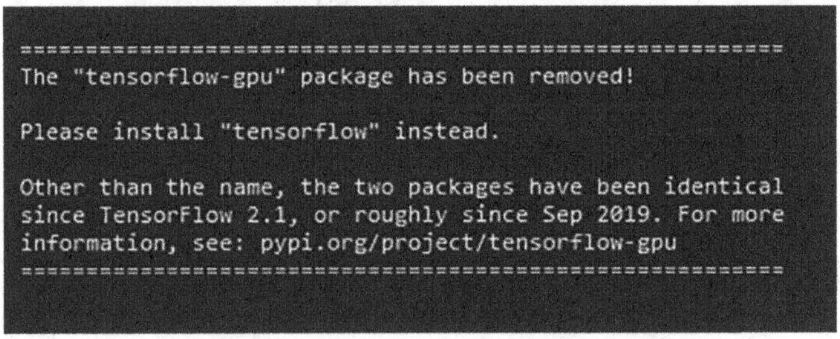

```
================================================================
The "tensorflow-gpu" package has been removed!

Please install "tensorflow" instead.

Other than the name, the two packages have been identical
since TensorFlow 2.1, or roughly since Sep 2019. For more
information, see: pypi.org/project/tensorflow-gpu
================================================================
```

Figure 4-16. *Deprecated packages can cause problems in workflows*

3. Run the docker image with GPU support using docker run.

4. Finally in a Jupyter notebook in your lab environment, you can check install using

    ```
    import tensorflow as tf tf.config.list_physical_
    devices('GPU').
    ```

You should now be able to run GPU accelerated code in TensorFlow without additional code changes. In the following chapter, we'll look at a detailed example of GPU accelerated training using the MNIST data set and the GPU enabled lab environment we just built (optionally, you can use Google Colab if you don't have a physical GPU device). Okay, so we have talked about using hardware to accelerate training, but what about processing large amounts of data? In the next section, we will look at how we can coordinate the processing of massive amounts of data using multiple processors in parallel. These types of databases are called massively parallel analytic databases or MPP.

Databases for Data Science

The distinction between an analytical system and a transactional system is an important one in data science. Transactional systems, also called "online" or operational systems, are designed to handle a large number of very small transactions (e.g., update one row in a table based on a primary key). These types of systems may support business processes like point of sales systems or other operationally critical parts of the business where speed and precision are nonnegotiable.

In contrast, analytical systems are designed to support offline workloads, large volumes of data, and queries over the entire historical data set. These analytical systems are usually implemented as a MPP (massively parallel processing) database. The types of queries that these databases can handle include large CTEs (common table expressions), window analytical functions, and range joins for point in time data sets. Snowflake is one such choice of MPP database. An example of a complex query that uses common table expressions and window analytic functions is given in Listing 4-1.

Listing 4-1. A Common table expression with a window analytic function

```
-- Use a common table expression to deduplicate data
WITH cte AS (
  SELECT id, component, date, value, ROW_NUMBER() OVER
  (PARTITION BY id, component ORDER BY date DESC) AS rn
  FROM sensor_data
)
SELECT id, component, date, value
FROM cte
WHERE rn = 1;
```

In this example, we first create a mock sensor data table "sensor_ data" with four columns: id, component, date, and value. We then insert some sample data into this table.

Next, we define a common table expression (CTE) and give it a name. This code is available as part of Chapter 4 (see example_deduplicate_data. sql). You can optionally run it by creating a cloud service account on Snowflake. Similar to the Databricks community edition, you can get a free trial using the self-service form on the Snowflake website; however, this is optional and the query will likely run with some modification on most MPP database that supports ANSI SQL since window analytic functions are a part of the standard since 2003. Let's break this query apart into its component pieces to understand how to write a query:

The SELECT statement is used to select all four columns from the sensor_data table, and we use ROW_NUMBER() window analytic function to assign a unique row number for each row. The PARTITION BY clause ensures that each row number gets reset for each combination of id and component.

Finally, the other SELECT statement selects four columns from our CTE but filters only on rows where the row number is equal to 1. This has the effect of de-duplicating our sensor_data table. You may find queries like this or even more complex CTEs in typical data science workflows which, for example, in this case may be used to de-duplicate a data set prior to running a train-test-split algorithm, avoiding data leakage. Of course, this is only a simple example.

Snowflake (a type of MPP database) SQL supports a wide range of window analytic and statistical functions for data science tasks such as ranking rows within a partition, calculating running totals, and finding the percentiles of a set of values.

Here are some examples of the types of functions that are commonly used in feature engineering.

- *Ranking functions:* ROW_NUMBER(), RANK(), DENSE_RANK()

- *Aggregate functions:* SUM(), AVG(), MIN(), MAX(), COUNT()

- *Lead and lag functions:* LEAD(), LAG()

- *Percentile functions:* PERCENT_RANK(), PERCENTILE_CONT(), PERCENTILE_DISC()

- *Cumulative distribution functions:* CUME_DIST()

- *Window frame functions:* ROWS BETWEEN, RANGE BETWEEN

- *Date and time functions:* DATE_TRUNC(), DATE_PART(), DATEDIFF()

- *String functions:* CONCAT(), SUBSTRING(), REGEXP_REPLACE()

In the next section, we will briefly detail patterns for enterprise grade database projects so we can get familiar with common architectural patterns.

Patterns for Enterprise Grade Projects

Data lake: A data lake is centralized repository, typically separated into bronze, silver, and gold layers (called the medallion architecture[2]). The central repository allows you to store both structured and unstructured data, contrasting with a traditional relational database. If you use a cloud storage account like blob storage or s3 buckets, the bronze, silver, and gold layers can map to containers or buckets where you can administer permissions and assign users or service principals access to each container or bucket. The bronze layer is the ingestion layer and should be as close to the raw data sources as possible (you can, e.g., organize raw data sources in folders, but it is important to have a consistent naming convention across the data lake). The silver layer is most important for data science and contains cleaned and conformed data that is still close enough to the source that it can be used for predictive modeling and other data science activities. The gold layer is used for enterprise grade reporting and should contain business-level aggregates.

Data warehouses: Data warehouses are an older pattern and are a centralized repository of data. Data can be integrated from a variety of sources and is loaded using ELT or ETL patterns. The data warehouse can contain dimensional data (slowly changing dimensions) and other tables. This architectural pattern is not well-suited for data science workflows which require flexibility and have to handle schema drift but can be used as a valuable data source for many projects.

Data mesh: A data mesh is a decentralized approach to building data stores that uses self-service design and borrows from domain-oriented design and software development practices. Each domain is responsible for their own data sources, requiring a shift in responsibility, while the data platform team provides a domain-agnostic data platform.

---

[2]www.databricks.com/glossary/medallion-architecture

Databases are not only used for feature stores (to organize features for model training) but also for model versioning and artifact storage; in fact, MLFlow also uses a database. Databases are also used for logging and monitoring. This is an important fact that is often overlooked in MLOps. In the next section, we will look at No-SQL databases and how we leverage meta-data in our data science workflows to adapt to change.

No-SQL Databases and Metastores

Relational databases can represent structured data in tables with relationships (foreign keys) between tables. However, not all data can be forced into this pattern. Some data, especially web data (JSON and XML), are semi-structured having nested hierarchies, and text-based data common in NLP problems are unstructured. There is also binary data (common when you have to deal with encrypted columns), and having to store, process, and define the relationships between structured, semi-structured, and unstructured data can be cumbersome and inefficient in a relationship database creating technical complexity. This complexity is compounded by schema evolution common to data science workflows. Hence, there is a need for an efficient way to represent, store, and process semi-structured and unstructured data while meeting nonfunctional requirements like availability, consistency, and other criteria important to the data model. In this section, we will introduce both No-SQL and relational databases that you can use to build data models and meet nonfunctional requirements without having to pigeonhole your solution into a relational database.

- *Cassandra:* Cassandra is a No-SQL distributed database that supports high availability and scalability which makes it ideal as an online feature store.

- *Hive:* Hive is a distributed, fault-tolerant data warehouse system for data analytics at scale. Essentially a data warehouse is a central store of data that you can run queries against. Behind the scenes, these SQL queries are converted into MapReduce jobs, so Hive is an abstraction over MapReduce and is not itself a database.

  - *Hive metastore:* Hive metastore is a component you can add to your feature store. It contains names about features such as names of features, data types (called a "schema"). It is also a commonly used component in cloud services like Databricks delta tables, so even if you aren't building your own feature store directly, you should have some knowledge of this important piece of data infrastructure.

Relational Databases

- *Postgresql:* Postgres is a relational database system that can support gigabytes, terabytes, and even petabytes of data. We can also configure PostGres to work with Hive metastore. In Feast (in version greater than 0.21.0), Postgres is supported as a registry, online and offline feature store.

Introduction to Container Orchestration

We learned about Docker and even created a Dockerfile which was a series of instructions used to build an image. The image, a binary containing layers of software, could be run like a lightweight virtual machine on our host operating system. But what if we have to run multiple services, each with their own Docker images? For example, we might have a service that hosts a Jupyter notebook where we type in our Python code, but we might want to have another service for storing data in a database and have our notebook interact with this database.

One subtlety you will encounter is networking. How can we get these two services to "talk" to each other and create the network infrastructure to support this communication between services?

Also since containers are ephemeral in nature, how do we spin these services up when we need them and spin them down when they're no longer needed while persisting the data we need? This is what container orchestration deals with, and the standard tool for orchestrating containers is Docker Compose.

We'll be using Docker Compose in the next chapter to set up MLFlow and get it to "talk" to our Jupyter lab which we will need to set up experiment tracking for our training pipeline. You can look at the docker-compose.yml file included with this chapter (but don't run any commands just yet). Figure 4-17 shows an example YML file.

```
docker-compose.yml
 1    version: '3.7'
 2
 3    services:
 4      jupyter:
 5        build:
 6          context: .
 7          dockerfile: ./lab/Dockerfile
 8        container_name: jupyter
 9        restart: always
10        ports:
11          - "8080:8888"
12        environment:
13          - JUPYTER_ENABLE_LAB=yes
14        command: start.sh jupyter lab --LabApp.token=''
15        volumes:
16          - ./notebooks:/home/jovyan/work
17        user: $UID:$GID
18
```

Figure 4-17. *A Docker Compose YML file*

Commands for Managing Services in Docker Compose

Container orchestration is a large topic, and as we mentioned, you will be using it in the next chapter to set up MLFlow and build a training pipeline. We'll cover MLFlow in depth, but Docker Compose has commands for managing the entire lifecycle of services and applications.

Here are some important commands useful for managing services:

- *Start services:* docker-compose up

- *Stop services:* docker-compose down

- *Start a specific service:* docker-compose up <service name>

- *Check the version of Docker Compose:* docker-compose-v

- *Build all services:* docker-compose up --build

While Docker Compose simplifies the process of creating services, you still need to define multi-container applications in a single file. Imagine a scenario where you have infrastructure that spans across different cloud providers or is multitenant in nature. This kind of multitenancy contrasts with multi instance architectures and having a tool that can completely describe infrastructure as code can help with the complexity in these environments.

- **Infrastructure as Code**

    Infrastructure as code (IaC) is a DevOps methodology for defining and deploying infrastructure as if it were source code. We have already seen an example of this when we spun up our data science lab environment by defining the image, binaries, and runtime needed inside the Dockerfile. Since the Dockerfile itself is a series of instructions that can be source controlled and treated like any other source code, we can use it to generate the exact same environment every time we build the image and run the container from the image. The ability to have the same environment each time is called *reproducibility* and is an essential component for data science because experiments need to be reproducible.

It's worth mentioning that there are specific tools and specialties within DevOps for managing infrastructure as code. One tool that is widely used in industry is Terraform. Terraform is an open source infrastructure-as-code tool for provisioning and managing cloud infrastructure such as Databricks. It works with multiple cloud providers and allows MLOps professionals to codify infrastructure in source code that describes the desired end state of the system. An example configuration file is given in the following, but these files can get very complex, and you can use Terraform and similar tools to configure and manage notebooks, clusters, and jobs within Databricks. Figure 4-18 shows a very simple example of infrastructure as code in Terraform.

```
                                                              Copy
resource "databricks_repo" "this" {
  url = "https://github.com/user/demo.git"
}
```

Figure 4-18. Infrastructure can be described as code

Making Technical Decisions

We've come a long way in this chapter from introducing Docker, applying what we learned to create our own data science lab environment complete with Jupyter notebook, and getting our hands dirty with Feast, creating our own feature store from an IoT data set.

We've also talked about the philosophy of having infrastructure as code and why it's important for the reproducibility of data science experiments. The final piece of the puzzle is how we can use our knowledge of infrastructure to make better technical decisions. Here are a few key points you should consider when making decisions around infrastructure in your own projects:

- Solve problems using a divide-and-conquer strategy, breaking services and parts of applications into functional components.

- Ask yourself if there is a cloud service or a docker container you might want to use for each functional component in your system.

- Understand the performance requirements for your workload. Do you need a lot of memory? Or do you need dedicated CPUs and GPUs for model training? Understanding the hardware requirements for different models can help you decide.

- Run code profilers on your code to identify bottlenecks in a data-driven way. A great profiler that comes with Python is cProfiler. It's often not enough to "guess"; you should strive to make data-driven decisions by *performance testing* your code.

- Strive to make your experiments reproducible, and deploying by using Docker and adopting a mentality of infrastructure as code can help to manage changes and different versions of infrastructure.

- Decide between PaaS (Platform as a Service) and Infrastructure as a Service. Sometimes, spinning up your own dedicated server and worrying about upgrades, updates, and security batches can be overkilled when a good PaaS meets infrastructure requirements.

Summary

In this chapter, we've learned the fundamentals of infrastructure for MLOps. We've covered sufficient prerequisites for understanding containerization, cloud services, hardware accelerated training, and container orchestration and how we can use our knowledge to become better technical decision-makers on data science projects. At this point, you should understand what a container is and how to build containers and be able to define what container orchestration means and why it is useful for MLOps. Although this chapter covered a lot of ground and you're not expected to know everything about containerization, we hope this chapter has peaked your curiosity as we start to build on these fundamentals and apply what we learned to some real data science problems in the coming chapters. Here is a summary of the topics we've covered:

- Containerization for Data Scientists

- Hardware Accelerated Training

- Feature Store Pattern and Feast

- GPU Accelerated Training

- MPP Databases for Data Science

- Introduction to Container Orchestration

- Cloud Services and Infrastructure as Code

- Making Technical Decisions

CHAPTER 5

Building Training Pipelines

In this chapter, you will build your own toolkit for model training. We will start by discussing the training and how it relates to the other stages of the MLOps lifecycle including the previous stage feature engineering. We'll consider several different problems that make this part of the lifecycle challenging such as identifying runtime bottlenecks, managing features and schema drift, setting up infrastructure for reproducible experiment tracking, and how to store and version the model once it's trained. We'll also look at logging metrics, parameters, and other artifacts and discuss how we can keep the model, code, and data in sync. Now, let's start by talking at defining the general problem of building training pipelines.

Pipelines for Model Training

Building pipelines are a critical part of the MLOps lifecycle and arguably the most essential part of the development and deployment of machine learning systems since training the model is the process that allows you to determine what combination of weights, biases, and other parameters best fit your training data. If model training is done correctly, meaning we've correctly minimized a cost function that maps to our business problem,

© Dayne Sorvisto 2023
D. Sorvisto, *MLOps Lifecycle Toolkit*, https://doi.org/10.1007/978-1-4842-9642-4_5

then our end result of this process will be a model capable of generalizing beyond our training set, to unseen data, making predictions that can be actioned upon by decision-makers.

In this chapter, we will take a step back looking at model training instead as a process. We'll learn how to represent this process in a natural way as a machine learning pipeline. We'll also consider what can go wrong in this critical step of the MLOps lifecycle including what happens when we can't train our model in a reasonable amount of time, what happens when our model doesn't generalize, and how we can bring transparency and reproducibility into the training process by setting up experiment tracking. We'll also consider a part of model training that is often overlooked: model explainability and bias elimination. Let's look at some high level steps you might encounter in a training pipeline.

ELT and Loading Training Data

Model training typically occurs after you've already collected your data and, preferably, you have a feature engineering pipeline in place to refresh the data. This is a complicated step. We looked at some of the data infrastructure you can use for building feature stores in the last chapter such as relational databases, massively parallel databases, and Feast and Databricks, but if you've ever had to build an ETL (extract, transform, and load) or ELT (extract, load, and transform) pipeline, you know that it involves setting up connection strings to databases and writing SQL queries to read data, transform it, and load it into a target database. You need to set up tables, handle schema drift, and decide what tools to use for scheduling your pipeline. This is a large topic within data engineering, and we can't possibly cover every detail of this process, but we can provide you with knowledge of a few tools for building feature engineering pipelines:

Tools for Building ELT Pipelines

Data science projects need a solid foundation of data engineering in order to support the feature engineering process. Challenges exist around the part of the MLOps lifecycle between when data is collected and when that data is cleansed, transformed, and stored for downstream model training tasks. The steps that go into this are commonly called ELT or ETL (extract, transform, load), and there are data specialists that focus on this area alone. ELT is the preferred choice for data science teams since we want to first extract and then load the data in a database. Once the data is loaded, the data science team is free to transform the data as they wish without having to specify the transformation beforehand. With the ETL pattern, you need to transform the data on the fly before it is loaded which can become difficult. In the ELT pattern, the data science team can select the features that they want with data already loaded in the database and run experiments on raw data or iterate toward the feature engineering required for building the models. We also want to separate our extract, transform, and load steps, and we need a tool that is capable of passing data between steps and comes with monitoring, scheduling, logging, and ability to create parameterized pipelines. More specifically for data science, we also want to support both Python and SQL in our pipeline. Let's take a look at a few of these tools for ELT in data science.

Airflow v2: Airflow (version 2) provides an abstraction called a DAG (directed acyclic graph) where you can build pipelines in Python, specify dependencies between steps (e.g., read data, transform data, and load data), have steps run in parallel (this is why we use a DAG to represent the pipeline as opposed to a more linear data structure), and provide a convenient web interface for monitoring and scheduling pipeline runs. You will want to use at least version 2 of Airflow since version 1 requires you pass data between steps using xargs. You can build full end-to-end training pipelines in Airflow locally, but when it comes time to deploy your models in production (we'll talk about this in depth in a coming chapter), you might want to set up Airflow as a cloud service. There are a couple options available for production Airflow workflows in the cloud such as Astronomer or Google Cloud Composer (based on Google Kubernetes Engine).

The other much more difficult option is to deploy your own Airflow instance to Kubernetes. This option is not recommended for the data scientist that wants to manage their own end-to-end lifecycle because setting up your own Airflow instance in production on Kubernetes does require knowledge of infrastructure and there are many cloud services available that provide high availability and reliability, so if you are managing the entire lifecycle end to end, it's recommended you choose a cloud platform like Astrologer provides Airflow as a service, so you don't have to deal with the low level details required to configure Airflow.

Azure Data Factory and AWS Glue

If you've worked on ETL or ELT pipelines in the cloud before, you've probably heard of AWS Glue or Azure Data Factory depending on your choice of cloud provider. Both of these options can be used especially in combination with PySpark since Azure Data Factory has an "activity" (pipeline step) for running notebook Databricks, and AWS Glue can also run PySpark for extract, transform, and load steps. One thing to consider when choosing an ELT tool is which dialect of Python is supported since for data science, you will likely be writing your extract, load, and transform steps in a combination of Python and SQL. Although this isn't a hard requirement, if the rest of your workflow is written in Python such as the data wrangling or feature engineering steps, you would need to figure out how to operationalize this code as part of your pipeline, and if you choose a low code or visual ELT tool that doesn't support Python, you will have to have the additional step of translating your entire workflow which may not be possible especially if you have complicated statistical functions. This also leads to the second consideration for choosing an ELT tool for feature engineering pipelines: Does the tool support statistical functions required by your workflow? If the tool supports Python scripts, then the answer is probably yes, but you should still consider what kind of packages can be installed. The same applies if your data science workflow is in another language other than Python, for example, Julia or R, and you need to consider how much community support there is for your language, and using a language that isn't widely used may restrict the options you have for building your pipeline.

Another option for ELT is choosing a tool that supports the entire machine learning lifecycle end to end such as Databricks. The advantage of having a single platform is reduced effort and fewer integrations compared to a component-based system, but you still need to consider many questions such as how you're going to organize your feature engineering pipeline, what does the folder structure look like? Where will the ELT scripts live? How can I add Git integration and set up jobs to run these scripts to refresh and update data required for the model?

The last piece of advice for this section is to have as much explicit logging and error handling as possible baked into your pipeline. Although as data scientists, we might be more focused on accuracy of our scripts, when you go to deploy your pipeline to production and it breaks, you will wish you had more information in the logs and spent more time handling errors in a graceful way. Adding some basic retry logic, try-except blocks, and basic logging can go a long way to making your feature engineering pipelines robust and reliable.

Using Production Data in Training Pipeline

It goes without saying that you need production data in your training pipeline. It makes very little sense to train a model if the data is not accurate and up to date. This may pose some challenges for teams that have strict security protocols. You may need to communicate your need for production data and the business need for requiring daily or real-time refreshes of this data. For most workflows, daily frequency should be adequate, but know that if you require low latency data refreshes, it may require additional infrastructure and code changes to support this. You may have to consider using an event driven architecture rather than a batch ELT pipeline.

Preprocessing the Data

Okay so you have your ELT pipeline, and you've decided how you're going to refresh the data and the frequency of updates and have chosen your feature store where your features will live. Your pipeline runs daily. You have code to read this data into a dataframe, maybe a Pandas dataframe or a PySpark dataframe if you're working on a structured data set, or maybe you use some other libraries like Spacy for processing text based data in an unstructured format.

The point is, whether the data is structured or unstructured, the shape, volume, quality of the data, and type of machine learning problem determine how it will be processed. There are many variables here so your preprocessing steps may be different.

What matters is how you are going to translate your assumptions about your model into code. Your data may have many missing values, and your model might require a value so you will have a preprocessing step to handle missing values. You may be solving a classification problem and found your data set is imbalanced, so you may have another step that resamples your data to handle this. Other steps might include scaling the data and getting the data in a shape the model expects. Take a look at Listing 5-1 for an example.

Listing 5-1. A code snippet showing preprocessing steps

```
from sklearn.preprocessing import StandardScaler
from sklearn.model_selection import train_test_split

X = df.drop('label', axis=1)
y = df['label'']
```

```
# train test split
X_train, X_test, y_train, y_test = train_test_split(X, y, test_
size=0.3)

scaler = StandardScaler()
# fit transform on X_train
X_train = scaler.fit_transform(X_train)

# transform X_test
scaler.transform(X_test
```

So how do we handle all of these preprocessing steps? It's really important to keep them all in sync, and this is why you need to use a pipeline. Although your ELT pipeline should be deployed using something like Airflow, for the complex sequence of transforms, most machine learning frameworks have a concept of a pipeline you can leverage for the transformations. For example, in sklearn, you can import pipeline from sklearn;pipeline as shown in Listing 5-2.

Listing 5-2. Importing sklearn's pipeline class

```
from sklearn.pipeine import Pipeline
```

Handling Missing Values

Missing values in data can have many root causes. It is important to assess the reason why data is missing before building your training pipeline. Why? The reason is simple: Missing values mean you do not have all of the information available for prediction, but it could also indicate a problem with the data generating process itself, human error, or inaccurate data.

Missing at random (MAR) is a term used in data analysis and statistics to indicate that the missing data can be predicted from other observed values in the data set. While not completely random, data that is missing at random or MAR can be handled using techniques like multiple imputation and other model based approaches to predict the missing value, so it's important to understand if the data qualifies as MAR or not. An example in finance would be a stock market forecast. Let's suppose you are tasked to build an LSTM model to forecast the price of a stock. You notice data is missing. Upon further investigation, you realize the missing data is correlated with another variable that indicates the stock market was closed or it was a holiday. Knowing these two indicator variables can be used to predict if the value was missing, so we say the price of the stock is missing at random. We might consider multiple imputation as a technique in our preprocessing steps to replace this missing value, or maybe it makes more sense to drop these values entirely from our model if the loss of data won't impact the accuracy of our forecast too much.

In addition to MAR, there is also MCAR (missing completely at random) and MNAR (missing not at random). With MCAR we assume that the missing data is unrelated (both to covariate and response variables). Both MCAR and MAR are ignorable; however, MNAR is not ignorable meaning the pattern of missing values is related to other variables in the data set. An example of MNAR would be an insurance survey where respondents fail to report their health status when they have a health status that might impact the insurance premium.

Knowing When to Scale Your Training Data

Scaling is applied when we have different units and scales in our training data and we want to make unbiased comparisons. Since some machine learning models are sensitive to scale, knowing when to include scaling in your training pipeline is important. Some guidelines for knowing when to scale your training data are as follows:

1. Do variables have different units, for example, kilograms and miles?

2. Are you using regularization techniques such as
    Ridge or Lasso? You should scale your data so that
    the regularization penalty is applied fairly, or you
    may have a situation where variables with larger
    ranges are penalized more than variables with
    smaller ranges.

3. Are you using a clustering algorithm that is distance
    based? Euclidean distance is highly sensitive to outliers
    and scale, so scaling your training data is necessary to
    avoid some variables dominating the computation

A general rule of thumb is to apply scaling to the numerical variables
in your data since from an MLOPs perspective, even if the model does not
require it, you can improve the numerical stability and efficiency. Now that
we've covered some of the preprocessing steps you might encounter in a
training pipeline, let's talk about a problem you will face when features
change: schema drift.

Understanding Schema Drift

Let us suppose you are a data scientist at a large financial institution.
You are creating a model to predict customer churn but need to consider
demographic and macroeconomic data. You recently were asked to add
another variable to you model: the pricing and subscription type for each
level of customer. You have five variables to add, one for each subscription
type; however, you will have to adjust your entire training pipeline to
accommodate them. This situation is called schema drift.

There are many ways to deal with schema drift, but as a general rule,
you should build your training pipeline in a way that is flexible enough
to accommodate future changes in variables since they will inevitably
happen. This might be as simple as altering a table to add a new column
or as complex as dynamically generating SQL including variable names

and data types, creating the table on the fly as part of the training pipeline. How you deal with schema drift is up to you, and some frameworks like Databricks provide options such as the "mergeSchema" option when writing to delta tables, so if you are using an end-to-end machine learning platform or feature store, you should consult the documentation to check if there is anything related to schema drift before building out a mechanism yourself.

Feature Selection: To Automate or Not to Automate?

Feature selection is important from an MLOps perspective because it can dramatically reduce the size of your training data. If you are working on a prediction problem, you may want to discard variables that are not correlated with your target variable

An interesting question is how much of this process needs to be automated? Should your training pipeline automatically add drop variables as needed? This is likely very unsafe and could lead to disastrous consequences, for example, if someone adds a field by accident that contains PII (personally identifiable information), demographic data that violates regulatory constraints on the model or introduces data leakage into your model. In general, your training pipeline should be able to handle adding and removing features (schema drift), and you should monitor features for data and model drift, but having a human as part of the feature selection process, understanding the business implication of the features that go into your model is a safer bet than taking a completely hands-off approach.

Building the Model

Once we have preprocessed the data, the next step is to build the machine learning model. In our case, we will be using scikit-learn's logistic regression model. We can define the model and fit it to the training data, as shown in Listing 5-3:

Listing 5-3. Fitting a model in Sklearn

```
from sklearn.linear_model import LogisticRegression

model = LogisticRegression()
model.fit(X_train, y_train)
```

Evaluating the Model

Once we have built the model, the next step is to evaluate the model. We will use scikit-learn's accuracy_score function to calculate the accuracy of our model on the test data, as shown in Listing 5-4.

Listing 5-4. Evaluating the model

```
from sklearn.metrics import accuracy_score

y_pred = model.predict(X_test)
accuracy = accuracy_score(y_test, y_pred)

print('Accuracy:', accuracy)
```

It's important here that model evaluation can be a complex process that happens both during hyper-parameter tuning and after tuning when comparing the performance of tuned models. The former is commonly called the "inner validation loop" and is used on a subset of training data before being retested on another subset. The purpose of this procedure is to find the best hyper-parameters for the model. Once our model is tuned, we can compare tuned models, and this is called the "outer validation loop" where you may choose the best model. Optionally you may choose tore-train the best model on the combined train and test sets in hope of getting better generalization. In the lab, you will build a training pipeline and see how some of this process works in practice.

Automated Reporting

Recall when we defined the MLOps maturity model, we said the differentiator between the first and second phase was an automated training pipeline as well as automated reporting. While we will cover performance metrics and monitoring in the next chapter, it is imperative to have infrastructure set up for reporting during the training phase; otherwise, the model can be trained, and we need to make decisions on model performance. While many MLOps professionals consider reporting to be important, reporting on model performance, model drift, and feature drift or tying in the model output from the training phase with business KPIs is a difficult process. At minimum your team should have a dashboard so you can discuss the results of trained models with stakeholders. Examples include Power BI which can be deployed to a cloud service or rolling your own such as Dash in Python and hosting it on a web server in the cloud.

Batch Processing and Feature Stores

When training a model, you need to decide if you want to store all of the data in memory or process the data in a batch, updating the weights of the model for each batch. Although gradient descent is widely used, theoretically there are alternative methods for optimization, for example, Newton's method. However, one practical advantage of gradient descent based algorithms is it allows you to train the model in a distributed fashion, breaking up the training set into batches. You should be aware if there are batch versions of your algorithm available. Gradient descent usually refers to batch gradient descent which trains on the entire data set in one go, but there are two modes for batch training you can code yourself when using gradient descent as an optimization algorithm, and they're available in most deep learning frameworks: mini-batch and stochastic gradient descent.

Mini-Batch Gradient Descent:

Mini-batch gradient descent is a tweak to the regular gradient descent algorithm that allows you to train your model on batches of data. The size of these batches of data can be tuned to fit in memory but is usually a power of 2 such as 64 or 512 to align with computer memory. Since the gradients are calculated over the entire mini-batch, the model weights get updated for each batch. This kind of divide and conquer strategy has many performance advantages, the most obvious one is the ability to run your computations on a smaller subset of data rather than than the entire data set in one shot. This translates into reduced memory footprint and faster computations. The trade-off you should be aware of is, unlike the regular batch gradient descent on the full training data, with mini-batch gradient descent, you are only approximating the true gradient. For most cases, this is acceptable, and for larger scale machine learning projects, training on the entire data set for several thousand epochs may not be feasible.

Stochastic Gradient Descent

Stochastic gradient descent is another variation of the classical gradient descent algorithm, this time using a randomly selected sample point to compute the gradients. The gradient of the loss function is used to update the model weights for each randomly selected sample point. The advantage, like mini-batch gradient descent, is less memory usage and possibly faster convergence. However, since the points are randomly selected from the training data, we are still only approximating the true gradient, and this approximation can be particularly noisy. Therefore, stochastic gradient descent sometimes combines with mini-batch gradient descent, so the noise term gets averaged out over many samples, leading to a smoother approximation of the true gradient.

Implementing stochastic gradient descent in a deep learning framework like PyTorch is as simple as importing the SGD optimizer.[1]

Online Learning and Personalization

The definition of an online learning method is a scenario where you don't want to train on the entire data set but still have a need to update the weights of the model as new data flows in. This is intuitive if we understand Bayes' rule which provides one such mechanism for updating a probability distribution, but when it comes to classical machine learning, we need to use gradient descent.

Linear classifiers (SVM, logistic regression, linear regression, and so on) with SGD training may come with a function that can have online or mini-batch mode supporting delta data or both online and batch mode supporting both delta data and a full data set.

Linear estimators in Sklearn, for example, implement regularized linear models with stochastic gradient descent learning. In this case, the gradient of the loss is estimated for each data point that the weights are updated by computing the partial derivative (take a look at Chapter 2 for an example of working with partial derivatives and loss functions in code). An optimization that is used with stochastic gradient descent is decreasing the learning rate (impacting the model's ability to update its weights in response to new data) as well as scaling the training data with zero mean

---

[1] PyTorch SGD optimizer documentation https://pytorch.org/docs/stable/generated/torch.optim.SGD.html

and unit variance (we talked a bit about this earlier in the chapter). This method is called feature scaling and can improve the time it takes for the algorithm to converge (sometimes at the cost of changing the output as with an SVM).

While online learning through methodologies like the partial fit function can be used to reduce training time, you might ask first if it is necessary since you need to build mechanisms for incremental data load, support partial fit, fine-tune the last layer of the model, and freeze the rest or some other methodology for updating the weights on a small subset of data. This can complicate the training process, so unless there is a good reason for doing it, you might be better to consider hardware accelerated training or distributed training on a full data set. However, there are still great reasons to consider online learning other than performance, one being the ability to fine-tune a model and personalize the prediction and in such cases. In the next section, we'll take a look at another important aspect of model training: model explainability.

Shap Values and Explainability at Training Time

Machine learning algorithms are often viewed as "black boxes" that accept labels and input data and give some output. As we know from Chapter 2, most algorithms are not "black boxes"; they're built up from mathematical abstractions, and although these abstractions can be powerful, they're also low bias machines, ultimately trading interpretability for a higher variance (see bias-variance trade-off). Neural networks, especially deep neural networks consisting of several layers of neurons stacked, are an example with both high variance and low explainability.

Fortunately, solving the problem of model explainability has come a long way, and some of the most widely used tools are LIME and SHAP.

LIME: LIME is an acronym for local interpretable model-agnostic explanations. The goal of LIME is to show how each variable is used in the prediction. In order to achieve this, LIME perturbed each observation and fits a local model to these perturbations. The model's coefficients are then used to assign a weighting to the feature importance of each variable. This weighting can then be interpreted as how much each variable contributed to the prediction, allowing the data scientist to explain the model. LIME is typically more performant than SHAP.

SHAP: Shapley Additive Explanations or SHAP relies on the concept of a Shapley value, a mathematical construct that uniquely satisfies some theoretical properties (from cooperative game theory) of an ideal model-agnostic explainability algorithm. The Shapley values can be interpreted as how much each feature contributed to the prediction. An interesting consequence of using Shapley values, which are available for each observation, is you can use it for model fairness as well, for example, to estimate the demographic parity of features in your model. SHAP aims to approximate the model globally and gives more accurate and consistent results, whereas LIME, which approximates the model locally, is much faster.

Feedback Loops: Augmenting Training Pipelines with User Data

One way to evaluate the maturity of an MLOps solution is by asking if it can incorporate output of the model back into the model training process, creating a simple kind of feedback loop. Feedback loops are ubiquitous throughout engineering.

Hyper-parameter Tuning

The final section we need to cover is hyper-parameter tuning and how it relates to the entire training pipeline. We know that models have hyper-parameters which are exactly that extra parameters like depth of a tree, number of leaves for tree based models, regularization parameters, and many other parameters that can do a variety of things from preventing model-overfitting to changing the architecture or efficiency of the model.

If you look at a boosting model like gradient boosting machines, you may have many hyper-parameters, and knowing details about how the algorithm works, for example, does it grow leaf-wise or level-wise, is essential to using the model correctly and tuning it to your business problem.

How do we search through a search space? We face a problem of combinatorial explosion if we try to do a brute force approach. We might try random search which reduces the search space, but then we might randomly miss important parameters and not have the best model at the end. A common approach is to use Bayesian optimization for hyper-parameter search. With the Bayesian approach, the best combination of hyper-parameters is learned as the model is trained, and we can update our decisions on which parameters to search as the process progresses, leading to a much better chance of finding the best model.

How do we implement Bayesian hyper-parameter search? One library that you will likely run into is HyperOpt. One important point is that you can set up MLFlow's experiment tracking inside the Hyperopt objective function. This powerful combination of MLFlow and Hyperopt can be an invaluable piece of your workflow. If you run the lab, you can see this pattern implemented and integrate it into your own MLOps toolkit. In the next chapter, we'll build on top of this foundation and look at how we can leverage MLFlow for finding the "best model" to make predictions on unseen data and use this model and data as part of an inference pipeline, but first let's take a look at how hardware can help to accelerate the model training process.

So what can go wrong in the model training process? One problem is that of all the steps in the MLOps lifecycle, model training can take the most time to complete. In fact, it may never complete if we are only running on a CPU. Hardware acceleration, which as we discussed, refers to the process of using GPUs (or TPUs) to speed up the training process for machine learning models, reducing the runtime by parallelizing matrix and tensor operations in an efficient way. Fortunately, there is a straightforward way to know when you might need to consider hardware accelerated training; you only need to ask yourself two questions:

- How long does it take to train my model?

- Is this training time reasonable given the business requirements?

If the answer to the second question is no, you will need to use hardware accelerated training. For example, if your model takes 3 days to train on your laptop, this is probably not acceptable, but in some cases, it may be less obvious, and you will need to consider other variables like if you can run the training pipeline automatically outside of business hours; maybe a few hours of training time is acceptable to you. You might also consider how fast the data is growing and if you will need to share resources with additional models in the future. In this case, although a few hours of training time might technically be feasible in the short term, long term you will need to consider solutions like hardware accelerated training to speed up the process, so you can accommodate the scale that you need in terms of volume of data or number of models.

Model architecture is also a critical variable to consider since, for example, deep learning models are often very expensive to train, requiring hours or days to fine-tune the models. Long short term models (LSTMs), large language models, and many generative models like generative adversarial networks are best trained on a GPU, whereas if your problem only requires decision trees or linear regression models, you may have more leeway in what hardware you use.

Hardware Accelerated Training Lab

Open the Hardware Accelerated Training Jupyter notebook in your MLOPs toolkit Jupyter notebook lab environment (named Chapter_5_gpu_ accelerated_training_lab) or, optionally, in Google Colab if you do not have access to a GPU on your laptop.

In the example we've set up, you'll be using a slightly different deep learning framework than we've seen so far. You'll use this framework, TensorFlow, to train a simple neural network on the MNIST data set. As part of the training pipeline, you'll need to preprocess the data, define the model architecture, compile the model, and set up the models' optimizer and loss function.

The most important part of this lab is the line of code that sets the GPU explicitly, using GPU using the with tf.device("/GPU:0") context manager. This tells TensorFlow to use the first available GPU either on your laptop or in Google Colab to accelerate the training process.

Experimentation Tracking

Experiment Tracking software is a broad class of software used to collect, store, organize, analyze, and compare results of experiments across different metrics, models, and parameters.

Experiment tracking allows researchers and practitioners to better understand the cause and effect relationships that contribute to experimental outcomes, compare experiments to determine common factors that influence results, make complex decisions on how to improve models and metrics to improve experiment results, and also reproduce these results during model training.

Remember, model training is a process that involves data and code. We need a way to keep track of the different versions of code and models and what hyper-parameters, source code, and data went into this training process. If we don't log this information somewhere, we risk losing it, and this means we're not able to reproduce the results of the experiment, keep track of which experiments were actually successful or even worse, and answer even the most basic questions around why an experiment went wrong.

One tool that is arguably the gold standard when it comes to experiment tracking in machine learning is MLFlow. MLFlow allows you to store models; increment model version numbers, log metrics, parameters, source code, and other artifacts; and use these artifacts at a later stage such as in a model serving pipeline.

MLFlow is itself designed for end-to-end machine learning and can be used in several stages of the MLOps lifecycle from training to deployment. It can even be used in a research context when there is a need to quickly iterate on results ad hoc and keep track of experiments across different frameworks, significantly speeding up your research.

MLFlow Architecture and Components

Experiment tracking: This component is used for logging metrics, parameters, and artifacts under a single experiment. The tracking component comes with a Tracking API which you can use in your training pipeline to log these metrics, parameters, and artifacts during the training process. In practice, experiment tracking can be set up in the hyper-parameter tuning step and used in combination with other frameworks like HyperOpt.

Projects: The MLFlow projects component is less of a traditional software component and more of a format for packaging data science code, data, and configuration. You might use projects to increase the reproducibility of your experiments by keeping data, code, and config in sync and deploying code to the cloud.

Model registry: The model registry component enables data scientists to store models with a version number. Each time the training pipeline runs, you can increment this version number and subsequently use the model API to pull a specific version number from the registry for use in a downstream model serving or deployment pipeline.

Model serving: The MLFlow model serving component allows you to expose your trained models as a RESTFul API for real-time inference or batch inference modes. You can also deploy models to a number of different environments including Docker and Kubernetes. We will cover model deployment in a subsequent chapter, but this is a vast topic that requires the deployment of not just the model itself but additional monitoring, authentication, and infrastructure to support the way in which the model is used by the end user.

Now that we've covered the basic components of MLFlow, how do we begin to use it and set up our own experiment tracking framework? Although we've worked with services in our MLOps toolkit like Feast and Jupyter labs, standing up these services as stand-alone Docker images and Python packages, MLFlow is a complex service with multiple components. For example, the model registry may need to support models that can get quite large and require either an external artifact store. We'll be using an s3 bucket for this. Technically, since we want to keep everything running locally, we'll be using another service called MinIO which emulates an s3 bucket for us where we will store our models.

Fortunately, since the docker-compose file is built for you in the last chapter, you only need to run. Go to Chapter 5 folder and run docker-compose up (Do you remember what this command does?). Listing 5-5 shows how to build all services from scratch.

Listing 5-5. Running docker-compose up with –build option

```
docker-compose up -d --build
```

You should notice this command spun up several services for you including MinIO (our cloud storage emulator for model storage), our MLFlow server (we use a relational database called MySQL for experiment tracking), and MLFlow web server where we'll be able to view our experiments and models once they're registered. You'll also notice our Jupyter lab notebook exists as a service and can talk to MLFlow through the docker-compose network backbone.

Okay, that's a lot of technical details, but how do we actually start using these services? If you look at the docker-compose file, you'll notice we exposed several ports. MLFlow web server is running on port 5000, our MinIO cloud storage service runs at port 9000, and our Jupyter lab server runs on port 8080 like before. If you open a browser and enter localhost:8080, you'll be able to access your Jupyter lab. This is where we'll run all of our code in this chapter. Table 5-1 summarizes these services and where you can access them.

Table 5-1. *Table of service endpoints used in this chapter*

Service	Endpoint	Description	Credentials
MLFlow web service	localhost:5000	View all experiments and registered models	None
Cloud storage service	localhost:9000	You need to access this once to create an s3 bucket called "mlflow"	MinIO MinIO123
Jupyter lab	localhost:8080	Where we'll be building our training pipeline	None

You should open a browser and navigate to each of these services.

Now that we have built and evaluated our machine learning model, the final step is to track our experiments using MLFlow.

Next, we need to import the mlflow package on PyPi and set the name of our experiment (we've already installed Mlflow for you as part of the Jupyter lab service but it is available as a stand-alone Python package).

When you set an experiment, all runs are grouped under this experiment name (each time you run your notebook, you are executing code and this is what is referred to as a run). You might want to establish a naming convention for experiments. For example, if you use a notebook, you could use some combination of notebook name, model types, and other parameters that define your experiment. An example code in Listing 5-6 shows similar code to what you'll find in the lab.

Listing 5-6. Creating an experiment in MLFlow using mlflow package

```
import mlflow

# Start an MLFlow experiment
mlflow.set_experiment('logistic-regression-mlflow')

# Log the parameters and metrics
with mlflow.start_run():
    mlflow.log_param('model', 'LogisticRegression')
    mlflow.log_param('test_size', 0.3)
    mlflow.log_metric('train_loss', train_loss)

    # Log the model as an artifact
    mlflow.sklearn.log_model(logistic_model, 'logistic_model')
```

What is this code doing? First, we start an MLFlow experiment by calling the set_experiment function and passing in the name of our experiment. MLFlow also comes in different flavors. For example, we can use the MLFlow lightgbm flavor to log a lightgbm model or sklearn flavor to log a sklearn model like logistic regression (we'll build on our logistic regression example from previous chapters).

Knowing which flavor of model API we're using is important when we *deserialize* the model (a fancy way of saying, loading the model back from the model registry) as we want the predict_proba and predict methods to be available. However, it can be challenging to handle different types of models in a general way.

161

You now have enough background knowledge to start the lab where you will build an end-to-end training pipeline and log model to MLFlow.

MLFlow Lab: Building a Training Pipeline with MLFlow

If you haven't done this already, now is time to run docker-compose up in the Chapter 5 folder and confirm all services are started by navigating to the service endpoints in Figure 5-1.

Step 1. Navigate to MinIO cloud storage service located at localhost:9000 and enter the credentials provided in Figure 5-1.

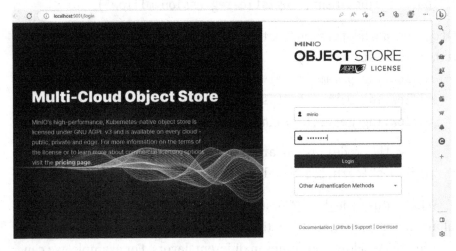

Figure 5-1. *MinIO Cloud Storage bucket*

Step 2. You need to create an s3 bucket where we'll store all of our models. Create a bucket called mlflow. If you're unfamiliar with cloud storage, you can think of this as an external drive, which we'll be referencing in our code. Figure 5-2 shows what the create bucket page looks like in MinIO.

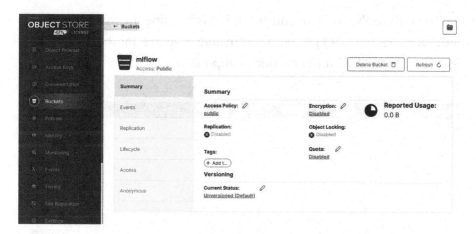

Figure 5-2. *Creating a bucket called mlflow in MinIO*

Step 3. Navigate to Jupyter lab service located at localhost:8080 in a browser, and import the notebook for Chapter_5_model_training_mlflow_lab. Read through all of the code first before running.

Step 4. Run all cells in the notebook, and navigate to the MLFlow web service located at localhost:5000. Confirm that you can see your experiment, runs, models, metrics, and parameters logged in the experiment tracking server. Figure 5-3 shows where MLFlow logs experiments.

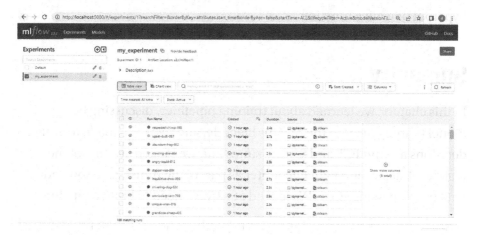

Figure 5-3. *MLFlow experiment component*

That is it! You've built an end-to-end training pipeline that trains a model and logs it to MLFlow, and you're able to search for the best run. Figure 5-4 shows the MLFlow model component.

Figure 5-4. *MLFlow model registry component*

Notice the last cell uses HyperOpt's hyper-parameter tuning framework to fine-tune the model. The important detail is how we define our search space and then set MLFlow's experiment tracking inside the hyperopt objective function.

Summary

In this chapter, we learned about training pipelines, discussing how model training fits into the MLOps lifecycle, after we have made technical decisions around ELT and feature stores and we looked at some of the high level steps you might encounter as part of the transformation and data preprocessing steps. We looked at why we need to build a pipeline and how we can make our pipelines more reliable and robust. We also discussed many of the technical aspects around setting up experiment tracking and hyper-parameter tuning. Here is a list of what you've learned up to this point.

- Tools for Building ELT Pipelines

- Preprocessing Data

- Hardware Accelerated Training

- Experimentation Tracking Using MLFlow

- Feature Stores and Batch Processing

- Shap Values and Explainability at Training Time

- Hyper-parameter Search

- Online Learning

- Setting Up an End-to-End Training Pipeline
  Using MLFlow

In the next chapter, we will build one some of the core ideas we learned to deploy models and build inference pipelines.

CHAPTER 6

Building Inference Pipelines

If you've made it this far, you've already created MLOps infrastructure, build a feature store, designed and built an end to end training pipeline complete with MLFlow experiment tracking for reproducibility and model storage in the MLFlow model registry, and tried monitoring and logging. It might seem like you're almost done; however, we're still missing a critical piece of the MLOps puzzle: Once you've trained your model, what do you do with it?

This is such a critical piece of the MLOps lifecycle that it's surprising so many data scientists leave the design and construction of the inference pipeline to the last minute or bury it away as a backlog item. The reality is, the inference pipeline is one of the most important parts of any stochastic system because it's where you will actually use your model to make a prediction. The success or failure of your model depends on how well stakeholders are able to use your model and action upon it to make business decisions; when they need it and without an understanding of this stage of the lifecycle, your project is doomed to failure. Not only that, but it's the inference pipeline where you will store the model output to incorporate feedback loops and add monitoring and data drift detection, so you can understand the output of your model and be able to analyze its results.

A lot can go wrong as well, and if you aren't aware of how to measure data drift and production-training skew, then your model may fail when it hits production data.

© Dayne Sorvisto 2023

D. Sorvisto, *MLOps Lifecycle Toolkit*, https://doi.org/10.1007/978-1-4842-9642-4_6

In this chapter, we will look at how we can reduce the negative consequences of production-training skew and monitor the output of our model to detect changes in problem definition or changes in the underlying distribution of features. We will also take a detailed look at performance considerations for real-time and batch inference pipelines and design an inference API capable of supporting multi-model deployments and pulling models from a central model repository similar to an architecture described in Figure 6-1.

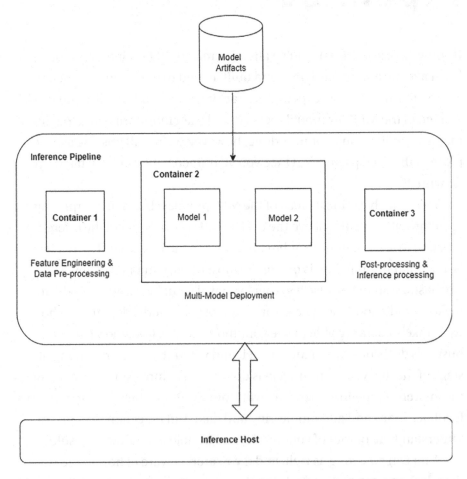

Figure 6-1. *Inference pipeline supporting multi-model deployment*

Reducing Production-Training Skew

Your model was trained on data that was carefully collected and curated for your specific problem. At this stage, you likely have a good idea of the distribution of features that go into your model, and you've translated certain assumptions about your model into the training pipeline, dealing with issues like imbalanced data and missing values.

But what happens when your model hits production and needs to make a prediction on unseen data? How can we guarantee that the new data follows the same distribution as the training data? How can we guarantee the integrity of the model output so that stakeholders can trust the output enough to action on the insights the model provides? This is where the concept of production-training skew comes into our vocabulary and starts to impact the technical decisions we make around model deployment.

Production-training skew can be formally defined as a difference in model performance during production and training phases. Performance here can mean the accuracy of the model itself (e.g., the unseen data has a different probability distribution than expected or can be caused by failing to handle certain edge cases in our training pipeline that crop up when we go to production).

It's worth noting that sometimes issues happen in production that are not anticipated even if we have a really good understanding of the assumptions of our data and models. For example, we might expect certain features to be available at inference time because they were available at training time, but some features might need to be computed on the fly and the data just may not be available.

In general, it is best practice to ensure your inference pipeline has safeguards in place to check our assumptions prior to using the model, and if features are not available, or if certain statistical assumptions are not met, we can have a kill switch in the inference to prevent the model from making an erroneous prediction.

169

This highlights an important difference between stochastic systems and traditional software systems because the consequences of actioning on a bad model output may be disastrous. As I've mentioned, stakeholders may lose trust in your model, or the model may be used as part of a decision process that impacts real people in a negative way. Therefore, it's not enough to fail gracefully or ensure our model always has an output; as an MLOps practitioner, you need to put model safety first and foremost and ensure that if critical assumptions are not met, then what went wrong gets logged and the inference pipeline fails.

Let's take a look at how we can set up monitoring and alerting to ensure the safety and integrity of our model.

Monitoring Infrastructure Used in Inference Pipelines

Although we have a firm grasp of infrastructure, we need to take a brief moment to talk about the type of infrastructure you will need to set up for monitoring your inference pipelines. There are various cloud-based monitoring services in all of the major cloud platforms such as Amazon CloudWatch, Azure Monitor, or Google Cloud Monitoring. These tools provide monitoring and alerting capabilities that can be integrated with data pipelines. Your organization may have their own monitoring infrastructure set up already in which case you should consider leveraging this instead of creating new services.

There are also specific monitoring tools for data pipelines and ELT frameworks; for example, AWS Glue and Airflow both have built-in monitoring, and you can use this to build your own custom data drift detection solution by creating a separate pipeline and setting up hooks that can talk to other infrastructure.

The difference between the data specific monitoring tools and the more general cloud monitoring tools is the more general cloud monitoring tools also can monitor resource utilization and you can use this to get a sense of where performance bottlenecks exist in your code. You may have to read the documentation for these cloud services and learn the SDK (software development kit), so you integrate these tools with your own code base. Whether you choose a stand-alone cloud monitoring service or leverage an existing one or one built-in with your ELT framework will depend on your project and the specific problem you're trying to solve.

Okay, so once you have made the technical decision on what type of monitoring service you want to use for your data drift and model drift detection, then we can talk about how you can implement monitoring in your inference pipeline and some of the challenges that you might encounter.

Monitoring Data and Model Drift

Monitoring is an essential part of nearly every IT operational system. It also happens to be one of the ways we can make data-driven decisions about our production models. Monitoring is a way of collecting data (strictly speaking, this is logging) and the capability of observing data over a period of time, for example, to check if certain conditions are met that are actionable. The action is usually called an alert.

It's important to realize that when working with monitoring systems, this data is collected in the form of logs, but the logs need not be centralized and are typically streamed via standard output and standard error and then consolidated using some logging service. Services include cloud services like Data Dog or Azure Monitoring or open source solutions like Ganglia, ElasticSearch.

In the context of machine learning and stochastic systems, monitoring means monitoring the model specifically for data drift and model drift and ensures the reliability and integrity of our model. We define these related terms.

Data drift: Data drift is related to the statistical properties and the probability distribution that underlies the features that go into training the model. When the underlying distributions of features shift in terms of mean, variance, skewness, or other statistical properties we can track, then it may invalidate assumptions we have made in the training pipeline and render model output invalid. A way to continuously monitor these statistical properties needs to be implemented.

How do you measure the difference between the distribution of features 6 months ago and at the present time? There are several ways to approach this, and one way is to measure the "distance" between two probability distributions such as with KL divergence or Mahalanobis distance. The important detail here is that we need to first measure a baseline and we compute this distance against the baseline, usually by defining a threshold value. If the divergence between our observed and baseline exceeds this threshold, then we can choose to send out an alert (e.g., an email to relevant stakeholders). It's important we actually send out an alert and build out the code to do this, for example, if your team uses Slack, you may consider building a slack bot to alert your data drift has occurred since important decisions need to be made on whether to retrain the model and understand the root cause of the shift.

Another approach to data drift is hypothesis testing. We can set the null hypothesis to the features that have not changed or come from a well-known distribution like the normal distribution if your data is normally distributed. One commonly employed hypothesis test is the Kolmogorov-Smirnov test where the null hypothesis is that the data comes from the normal distribution.

Once we have confirmed that data drift has occurred, we have to make a technical decision: Should we retrain our model? This is the first kind of feedback we can introduce into the training pipeline and is more sophisticated than the alternative which is periodically retraining on a set schedule (which may be a waste of resources if data drift has not occurred or is within the SLA threshold).

Model drift: Model drift is a slightly different concept than data drift and can indicate that the business problem has changed. It's important to define the business problem and the definitions of features as part of the feature engineering step so that you can validate if model drift has actually occurred once detected.

Detecting possible model drift is fairly straightforward but verifying it is not. In order to detect model drift, we only need to monitor the predicted values (or more generally, the output of the model) and compare them to the expected values over time. For example, if we have a multi-class classification problem, we might record the total number of predictions made for each class and the breakdown of our predictions by each class, counting the number of predictions made for each class. We could visualize this as a simple histogram where the bins are the classes in our model, and if we find this histogram changes too dramatically from the baseline (using since threshold we define for the specific problem), then we have data drift and suspected model drift (performance of our model may have degraded over time).

We may also keep track of accuracy and other performance metrics and keep track of the performance of our model over time and a baseline and confidence intervals if possible.

Once we have found that either the model output has changed or the model performance metrics are degrading, then we need to investigate if model drift is actually concept drift, meaning the business problem has shifted in some way. This may lead us not only to retrain the model but possibly to have to add new features, revise features, or even change the model and its assumptions entirely to match the new business problem.

173

In order to keep track of the model output, we need a reliable way to make predictions with our model (if the mechanism isn't reliable or at least as reliable as the model output, then we won't be able to tell when we model drift has occurred). Creating the API for inference is not only about user experience but also ensuring the accuracy and reliability of the model output. In the next section, we'll go over some of the considerations that go into designing a reliable inference API.

Designing Inference APIs

Okay so let's say we have the most reliable inference API, we trust the data and the output of our model, and our stakeholders and users trust the output. The next focus needs to be on performance. We've noted previously there are technical trade-offs between accuracy and model performance, and while we should always consider performance early, it's important not to sacrifice accuracy or fairness of the model for performance. On the other hand, if we don't consider scalability and optimize our inference pipeline for performance, then the output may be rendered completely invalid by the time the prediction is made (e.g., delivering the prediction the next day if there is a hard requirement on the latency of the system). Due to this performance-accuracy, trade-off in some sense performance is a two-sided problem in machine learning.

In the next section, we'll take a detailed look at what we mean by performance in the context of inference pipelines in terms of both scalability and latency but also accuracy and validity and some of the important performance metrics we should be tracking in our monitoring solution. We'll also discuss the important problem of alignment in data science and how it plays a role in deciding what performance metrics to track.

Comparing Models and Performance for Several Models

In Chapter 5, we looked at model training and talked about the model tuning step. On a real-world problem, you may have many different types of models that you need to compare. You may have to dynamically select the best model, and we need a way to compare models for a problem type to choose the best model we should use for model inference.

One approach is once our models are tuned, we evaluate their performance using k-fold cross-validation and by selecting the model that has the best performance, for example, accuracy of F-1 score. This "outer validation loop" may use cross-validation but is done after hyper-parameter tuning since we need to compare models once they are already tuned; it would make little sense to make a decision on what is the best model if we haven't even gone through the effort of fine-tuning the model.

Since we'll typically be working to solve one problem type like classification or regression or anomaly detection, there are common performance metrics we can use to decide objectively what the best model should be, and there needs to be code that can handle this part of the process. Let's take a detailed look at some of these metrics and performance considerations used for comparing models across problem types.

Performance Considerations

Model performance can refer to the accuracy and validity of our model or scalability, throughput and latency. In terms of accuracy and validity, there are many metrics, and it's important to choose the metrics that are aligned with the goals of the project and the business problem we want to solve.

Here are some examples; in this table, we try to break them down by type of problem to emphasize that we need to consider the *alignment of the model* with the goal. We call Table 6-1 the alignment table for data science.

Table 6-1. *Alignment table for data science*

Problem type	Metrics
Classification	Accuracy
Classification	Precision/recall
Classification	F1 score
Regression	RMSE/MAE
Recommendation	Precision at k
Recommendation	Recall at k
Clustering	Davies-Bouldin Index
Clustering	Silhouette distance
Anomaly detection	Area under curve (AUC)
All problem types listed	Cyclomatic complexity

Of course this is not an exhaustive list since we can't possibly list every problem type you may encounter. I hope it provides a good starting point for designing your inference pipeline. In the next section, we'll take a deep dive into the other side of performance: scalability and latency.

Scalability

How can your machine learning system handle increasing amounts of data? Typically, data collection, one of the first phases of the MLOps lifecycle, grows over time. Without further information, we don't know at what rate this data collection process grows, but even if we assume logarithmic growth, over time, we need to scale with the increasing data volume.

You might have heard the word scalability before in the context of machine learning, the ability of your system to adapt to changes in data volume. Actually, scalability goes in both directions; in fact, cloud services are often described as being "elastic," when you don't use them they should scale down and during peak periods of activity, they scale up.

What does it mean to scale up and down? We usually speak of horizontal scalability and vertical scalability.

Vertical scalability: Vertical scalability means we add additional memory, CPU, and GPUs or in the case of cloud services increase these physical resources on the virtual machine or compute we are running. By vertically scaling, we're adding more horsepower to a single worker machine, not adding new machines. This gets expensive after a while since as your memory or compute needs grow, at some point it is no longer feasible to upgrade the machine, and this is why for data science, we consider horizontally scaling workflows rather than vertically so we can leverage several inexpensive worker machines (often commodity hardware) to reach our compute and memory needs.

Horizontal scalability: Horizontal scalability means we add additional worker machines and consider the total compute (number of cores) or total memory of the entire cluster together. Usually, this comes with hidden complexity such as how we can network the machines together and shard the data across workers. Algorithms like map reduce are used to process big data sets across workers.

We mentioned in the previous chapter that we could use this horizontal scaling pattern for distributed training, but what about inference? When it comes to inference, we usually consider two types of patterns: batch mode inference and real-time inference.

Both of these patterns require different architectures and infrastructure but which one you choose depends on your particular use case (remember, we should always try and align technical decisions with our use case). Here is the definition of both batch inference and real-time inference.

Batch inference: Batch inference means we break our feature set into batches and use our model to run predictions on each batch. This type of pattern can be scaled out horizontally and also has the advantage of not requiring an API, load balancer, caching, API throttling, and other kinds of considerations that come with designing an API. If you only need to populate a table for a dashboard, for example, you might consider using batch inference. However, this pattern might be ill-suited for use cases requiring real-time or near real-time inference or on demand predictions.

Real-time inference: If your requirement is to have sub-second latency in your inference pipeline and event driven prediction or allowing the end user to make on demand predictions, then you may want to move away from batch mode and consider building an API. Your API can still be scaled horizontally using a load balancer, but you will need to set up additional infrastructure and an online feature store. If your requirement is sub-second latency, you may also need to use GPUs to make the prediction (or distributed pipelines). This is a complicated topic, and so in the next lab we'll discuss some of the components that go into building an inference API, and then you'll use MLFlow to register a model in production, pull it from the model registry, and explore how you might expose the model using an gRPC or RESTful API.

What Is a RESTful API?

A RESTful API is an interface between containers (or even remote servers) used to facilitate communication over the Internet (the communication protocol is called the HTTP protocol). RESTful APIs are created in frameworks like Flask to exchange data.

When an API endpoint is called either programmatically via a POST request (we can also do this manually using tools like Postman) or in the web browser (e.g., through a GET request), data (usually in the form of JSON) is serialized (converted to bytes) and sent across the Internet in a process called *marshaling*. The bytes are then converted back into artifacts like a model using a load function that is called deserialization. All of this happens transparently when you use a framework like Flask, and you can define endpoints (e.g., localhost:80/predict) which can be called either by other APIs or by applications that want to use your API (you could do this using Python's request library; you just need to specify the endpoint, the data, and if it's a POST or GET request you need to make).

APIs are one of the many ways to build inference pipelines that the user can interact with and are particularly suited as mentioned before for on demand use cases (you can just call the endpoint when you need it) or when you need a sophisticated application that uses your model (these applications are often built as microservices).

Although building a full API is beyond the scope of this book, it is worth being aware of a few technologies that are used in building large scale applications often called *microservices*.

What Is a Microservice?

A microservice architecture is an architectural pattern for software development that organizes applications (e.g., APIs) into collections of independent (in software development parlance, this is often called loosely coupled) components called services. We've already seen examples of services when we used docker-compose to build our Jupyter lab service and MLFlow service, but you can also build your own services. In practice, these services are self-contained API endpoints written in a framework like Flask. Since the services are loosely coupled, they will need to talk to each other by sending data in the form of *messages*. These messages are usually sent by calling an API. Since the services are loosely coupled docker containers, they can be scaled horizontally by adding more containers and distributing the load over several containers using a component called a *load balancer*. Figure 6-2 shows a typical REST API endpoint for prediction in Flask. The function predict exposes a *route* called /predict and expects features to be passed in as JSON strings in the body of a POST request (a standard way HTTP endpoints accept data). The model is loaded or *deserialized* and then used to make a prediction on the input data. The prediction is then returned as a json string, called a response.

```python
import mlflow
from flask import Flask, jsonify, request
import pandas as pd


app = Flask(__name__)


@app.route("/predict", methods=["POST"])
def predict():
    # Get the JSON data from the request
    input_data = request.get_json()

    # Convert the JSON data to a pandas DataFrame
    input_df = pd.DataFrame.from_dict(input_data, orient="index").T

    # Logged_model is instantiated in cell above
    loaded_model = mlflow.sklearn.load_model(logged_model)

    # Make a prediction using the loaded model
    prediction = loaded_model.predict(input_df)

    # Convert the prediction to a JSON response
    response = {"prediction": prediction.tolist()}

    return jsonify(response)
```

Figure 6-2. *Flask API prediction endpoint*

If you want to learn more about Flask, it's recommended you read the Flask documentation or several books available on microservice architecture in Flask. For most data scientists, building a microservice would be overkill, require teams of developers, and if attempted yourself would open up your project to security vulnerabilities and problems with scalability. Remember, you need to have additional components like load balancers and container orchestration frameworks like Kubernetes (docker-compose was the container orchestration tool we learned, but Kubernetes requires specific expertise to use effectively).

However, the pattern in Figure 6-2 is called a *scoring script*, and if you choose a cloud service that supports model inference, it will likely have support for creating your own inference scoring scripts which allow you to wrap the prediction logic in a function and expose a REST endpoint. Examples of cloud services that support these scoring or inference scripts include Databricks, AWS SageMaker, and Azure Machine Learning Service and MLFlow. In the lab, we'll look at how to build your own inference API and some of the details involved in registering a model, loading a model, and exposing an API endpoint in enough detail that you will have the hands-on skill to work with many different cloud services.

Lab: Building an Inference API

In the hands-on lab, you will take the code we wrote for training and adapt it for model inference. First, let's look at some of the components of an inference API. You're encouraged to do the supplementary reading before continuing to the hands-on Jupyter notebook for this chapter (I've already included them when you start the Jupyter lab for convenience, but you should try to import them yourself.).

Step 1. Run docker-compose up to start MLFlow and Jupyter lab services.

Step 2. Open your Chapter 6 lab notebook called Chapter_6_model_inference_lab.

Step 3. Run all cells to register the model and increment the model version number.

Step 4. Pull the registered model from MLFlow model registry. An example is given in Figure 6-3.

```
# Set up MLFlow Logging
mlflow.set_tracking_uri("http://mlflow_server:5000")
mlflow.set_experiment("my_experiment")
client = MlflowClient()
```

```
# Get the best run
runs = client.search_runs(experiment_id, order_by=["metrics.accuracy DESC"], max_results=1)
best_run_id = runs[0].info.run_id

logged_model = f'runs:/{best_run_id}/logistic_model'

# Load model as a PyFuncModel.
loaded_model = mlflow.sklearn.load_model(logged_model)
```

Figure 6-3. *Pulling a model from MLFlow registry for use in an inference pipeline*

Step 5. Use the model to make a prediction.

Step 6. Open the deployment notebook (called Chapter_6_model_inference_api_lab) to see how MLFlow serve can be used to expose your model as an inference API.

Keeping Model Training and Inference Pipelines in Sync

In Chapter 1, we talked about how technical debt could build up in a data science project. In fact, data science projects have been described as the high interest credit card of technical debt. One subtle way projects can accumulate technical debt at the inference stage of the lifecycle is by not keeping the training and inference pipelines in sync.

The same features the model was trained on are required at time of prediction. So there we must generate those features somehow. It's convenient to think we could reuse the exact same code, but sometimes not all features will be available at prediction time and additional pipelines are necessary. A great example is a feature like customer tenure, very common in finance which technically changes every instant. This should be recomputed at inference time before being fed into the model especially if there's a large lag between when the features get refreshed and when the model is applied. Keeping training and inference pipelines in sync via shared libraries and the feature store pattern can shave off technical debt. While the problem of keeping pipelines in sync is a software engineering problem, some problems cannot be solved with software engineering since the root cause of the problem is a lack of data. One such example is the so-called "cold-start" problem.

The Cold-Start Problem

The cold-start problem is something we see in recommender systems but more generally when we're working with transactional data, for example, customer or product data in retail, finance, or insurance. The cold-start problem is a scenario where we don't have all of the history for a customer or we want to make predictions about something completely new. Since we may not have any information about a customer or product, our model won't be able to make a prediction without some adjustment. Collaborative filtering, an approach in machine learning to filter on "similar" customers or products where we do have information available, can be used to solve the cold-start problem and make predictions on completely new data points[1].

---

[1] In situations where there is no data, collaborative filtering may need to be supplemented with approaches such as content-based filtering.

Although we've covered quite a few things that could go wrong in our inference pipeline, we can't anticipate every possibility, and while continuous monitoring plays a crucial role in making our inference pipelines more robust, sometimes things go wrong, code gets handed off to other teams, and we need to dig deeper into the system for technical specifics. This is where documentation can be a lifesaver.

Documentation for Inference Pipelines

If you're a data scientist, you probably have copious amounts of documentation for features and statistical properties of those features, but one area where documentation may be lacking is around the assumptions that go into building an inference pipeline.

For example, do you have a naming convention for models in production? How about model versioning? Can you explain the process for updating a model or what to do if your inference pipeline breaks in production and your model isn't able to generate a prediction? All of these steps should be documented somewhere, usually in the form of a *run book*. It is also critical to have internal documentation such as a wiki that gets updated regularly. This documentation can be used for onboarding and hands-offs and to improve the quality of code and can save you when something inevitable breaks in production. Since documentation tends to only be used when things go wrong and stakeholders usually don't like reading large volumes of technical documentation, we also need a way of reporting performance metrics to stakeholders.

Reporting for Inference Pipelines

Reporting is another critical component of machine learning and in particular building the inference and training pipeline. With respect to the inference pipeline and model output, reporting is particularly important because the model output needs to be translated into business language using familiar terms that the stakeholders understand.

Since the ultimate purpose of the model was to solve a business problem, reporting could arguably be the most important piece of the puzzle as far as determining the value of your model.

Reporting can contribute to understanding the model, how the users are interacting with the model, and understanding areas of improvement and should be seen as a communication tool.

Reporting can take many forms from simple automated emails (remember we discussed one form of this for use in data drift and model drift monitoring) but also more sophisticated solutions like dashboards. Dashboards themselves should be viewed as operational systems that provide accurate data to an end user, bringing together multiple disparate data systems. Such systems may include the model output, feature store, user interaction with the model output (feedback loops), as well as other transactional or analytic database systems used by the business.

The type of dashboard you build depends on the business problem and how end users will ultimately interact with your models, but one dashboard that can add immediate value and prevent your model from ending up in the model graveyard is an explainability dashboard.

The ability to explain your model results with stakeholders is a crucial part of any data scientists' day to day role and information about model training, and what features are important when making a prediction (such as Shap value or lime) can serve as an invaluable communication tool. Some common use cases for reporting in MLOps include the following:

- Reporting on performance metrics

- Reporting on model explainability

- Reporting on model fairness for model bias reduction

- Reporting on feature importance

- Reporting on how the model output translates into key business KPIs

Reporting on how your model translates into key business KPIs is a critical exercise that should be taken into account from the beginning of the project before you even build the model, but keeping in mind that you need to translate this into a deliverable in the form of a dashboard at the end of the project can contribute to project planning help data scientists work backward from the dashboard through to the types of data and code needed to support the dashboard so a critical path for the project can be well-defined. Since data science projects have a tendency to suffer from lack of requirements or ambiguity, having a concrete deliverable in mind can reduce ambiguity and help to prioritize what is important in the project throughout the entire MLOps lifecycle.

Summary

We've come a long way in this chapter. We discussed how to build inference pipeline code examples along the way, and we actually built an inference pipeline with MLFlow and Sklearn in our hands-on lab. You should have a thorough understanding of the challenges that exist at this stage of the lifecycle from model monitoring, data drift, and model drift detection, aligning our problem to performance metrics and figuring out how to keep track of all of these performance metrics in a sane way. We discussed how to choose the best model when we have several different types of models. We gave some examples of performance metrics you may encounter in the real world for various problem types like anomaly detection, regression, and classification. We also discussed the importance of reporting, documentation, and keeping our training and inference pipelines in sync. Some of the core topics you should now have expertise include the following:

- Reducing Production-Training Skew

- Monitoring Data and Model Drift

- Designing Inference APIs

- Performance Considerations

In the next chapter, we'll look at the final stage of the MLOps lifecycle and formally define the lifecycle, taking a step back from the technical and developing a more holistic approach to MLOps.

CHAPTER 7

Deploying Stochastic Systems

If you've made it this far, you already have the skills to build a complete end to end data science system. Data science of course is more than machine learning and code which are really only tools, and to build end to end systems, we need to understand people, processes, and technology, so this chapter will take a step back and give you a bird's-eye view of the entire MLOps lifecycle, tying in what we've learned in previous chapters to formally define each stage. Once we have the lifecycle defined, we'll be able to analyze it to understand how we can reduce technical debt by considering the interactions between the various stages from data collection and data engineering through to model development and deployment. We'll cover some philosophical debates between model-centric vs. data-centric approaches to MLOps and look at how we can move toward continuous delivery, the ultimate litmus test for how much value your models are creating in production. We will also discuss how the rise of generative AI may impact data science development in general, build a CI/CD pipeline for our toolkit, and talk about how we can use pre-build cloud services to deploy your models. Without further ado, let's explore the stages of the ML lifecycle again and introduce the spiral ML lifecycle formally.

© Dayne Sorvisto 2023
D. Sorvisto, *MLOps Lifecycle Toolkit*, https://doi.org/10.1007/978-1-4842-9642-4_7

Introducing the Spiral MLOps Lifecycle

Although we hinted at the ML lifecycle throughout this book and even talked about the "spiral" MLOps lifecycle in Chapter 1 (shown in Figure 7-1), we lacked the context to really define the lifecycle completely and to understand the big picture from a holistic point of view. Although you might see the machine learning lifecycle or MLOps lifecycle (to me the difference between the two is that MLOps takes into account the business and IT environment the models live in), the reality is a lot messier. It's always been a pet peeve of mine that there's infographics used in data science that summarize complex ideas very concisely but don't map very well to real-world problems. Essentially these infographics are communication tools but not structures that can be mathematically defined or reasoned upon without a lot of imagination. Therefore, it's my goal to take the MLOps lifecycle infographic we saw in Chapter 1 to the level where you can actually recognize it in a real project or even adapt it to your own custom project since not all data science problems are the same across industries (maybe this is a kind of meta transfer learning problem in itself).

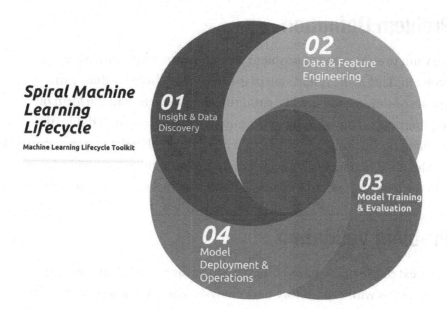

Spiral Machine Learning Lifecycle

Machine Learning Lifecycle Toolkit

Figure 7-1. *The spiral MLOps lifecycle*

So what is a lifecycle? In the context of biology, a lifecycle is a series of changes in the life of an organism. It's in itself a model for understanding change and a way of identifying the phases that come to define the organism over time.

Although MLOps is not a living organism, your IT environment is in many ways like a living, breathing organism. When we throw models and code into the mix, we legitimately have a kind of chaotic system, that although may not be technically living changes over time and is best understood by breaking it down into distinct phases.

The phases which include developing, deploying, and maintaining models can further be segmented into more granular stages that go into creating a useful machine learning model that solves a business problem.

Problem Definition

We want to start with the problem definition and requirements for the problem. This is the initial step of the lifecycle, where we define the initial conditions and involve gathering of information in the form of requirements. This is vitally important because if you can't define the problem or the requirements (a step often skipped in data science projects and justified in hand-wavy ways), then this ambiguity can trickle down into the user stories and individual developer tasks, creating chaos.

Problem Validation

The next phase of the lifecycle is validating the model. Some people confuse this with exploratory data analysis which is sometimes used as justification for finding a problem, but the goal should be to understand the problem better, what we're looking at, and the types of data sources available and validate whether or not we can solve the problem in the first phase. Problem validation is different than exploratory analysis though. Although this phase may be tedious, it saves a lot of time because it's relatively cheap to validate a problem but costly to implement a full solution that ends up missing the mark in the end.

Data Collection or Data Discovery

Once we have validated a problem, we can collect data. Collecting data is expensive. Even if you don't collect the data, there is still a lot of data discovery that has to take place. You may leverage metadata if you have a catalog built already, but if not, you may have to build the metadata catalog yourself and discover the name, variable names, data types, and statistical properties of the data.

Data Validation

At this stage, a decision needs to be made whether there is enough value in the discovered data sources or if you need to go back to the first stage to refine your problem. This is another example of a feedback loop or what we're calling a "spiral" since the process may be continuous and hopefully converge to a performance model at the end of the process.

Data Engineering

After data collection and data validation, the next step is data engineering. MLOps requires a solid data engineering foundation to support modeling activities, and this is not trivial. If you're the data engineering and MLOps practitioner on the team, then you might struggle to build this foundation before you're able to make use of MLOps.

You will set up feature stores, build feature pipelines, decide on a schedule for your pipeline (refresh rate), and ensure you're using production data sources. You have to decide which data sources are most valuable to operationalize. If the data sources are normalized (3NF or 2NF), you may have to join them together into a centralized repository.

At this stage, you may have a data architecture in mind such as a Data Warehouse or a Data Lake or Data Vault and build robust ELT pipelines. The goal of this phase is to have feature stores that are accessible, secure, and centralized and to ensure that there's enough data to support model training.

You may start feature engineering, building a library of features for model training, to support a variety of problem types.

If you are truly dealing with a prediction problem, you may only keep features with predictive value, but you may also have to deal with issues of multicollinearity and interpretability.

Model Training

Model development starts. You may start with a model baseline like simple linear regression or logistic regression; the simpler the model, the better the baseline. You might gradually increase the variance of the model. The boundary between the previous phase, data engineering, and model training will be blurred as you refine your model and require new features be added to the feature store and deal with schema drift and the curse of dimensionality. Eventually you'll have to create a way to reduce the number of features and may struggle to keep track of the entire library of features as business demand grows.

Since the lifecycle is a continuous process, after several iterations, the architecture of the model itself may change, and if enough data is available, you might consider deep learning at some point. This stage may start simple but also grow horizontally in terms of the number of models you need to support and the number of problem types and performance metrics that need to be tracked. As your MLOps process gets more advanced, the model training phase will eventually require MLFlow or similar experiment tracking software, hyper-parameter tuning frameworks like HyperOpt, hardware accelerated or distributed training, and eventually full training automation, registering your models in a model registry and having some kind of versioning system.

Diagnostic Plots and Model Retraining

Depending on the problem type, there are specific visual tools for evaluating models called diagnostic plots. For example, if you have a classification problem, you might consider plotting a decision surface for your model to evaluate its strengths and weaknesses. For a linear regression problem, you may be interested in plotting residuals vs. fitted values or some other variation to decide if it's a good or bad model.

Some of these plots may be used as diagnostic tools but not traditional monitoring tools which may not be able to accommodate images, so you could, for example, have a Jupyter notebook that's source controlled a part of the project and can generate these images on a schedule, for example, once a week, or another option is to build a separate monitoring dashboard using tools like Dash or PowerBI; the choice of reporting software really depends on your project and how you're comfortable creating the visuals, but it probably needs to support Python and libraries used like Pandas.

For model retraining, you can have more complex triggers such as if the distribution of features changes or if model performance over time is trending downward, but a simple solution to start is to retrain the model monthly. Note that these are two different types of triggers and both are needed to determine when to retrain the model since model performance can degrade over time but also the distribution of features themselves can change.

You should also consider how you write features to a table. For example, you might want to add a timestamp column and append features to a table so you have a complete history available for model retraining. These types of technical decisions around how frequently data needs to be updated, whether historical data needs to be maintained for model retraining, and how to operationalize diagnostic plots and other visuals are complex decisions that may require several team discussions.

Model Inference

In this phase, you'll select the best model, pulling the model from the model registry for use in an inference pipeline. You may decide to go through another round of cross-validation to evaluate the best model. You will need to have an inference pipeline that compiles your features and makes them available at prediction time. The runtime, model, and features will all need to be available at the same time for the model to make a prediction. Your inference pipeline may be as complicated as a full application or microservice or as simple as an API endpoint or batch inference pipeline depending on the requirements gathered during requirements gathering.

One commonality between model training and model inference is schema drift. Schema drift is also a factor in choosing a data architecture that can adapt to the demands of data science workloads since features that are used in both training and inference can change. The implication is that we either need to have complex code flows, updates, and frequent release cycles to accommodate changes in feature definition, or we need to create our tables dynamically using metadata. Since data types of each feature determine how the data is physically stored, changes in data type can impact our ability to store historical data required for model training. In the next section, I will talk about the various levels of schema drift.

The Various Levels of Schema Drift in Data Science

Schema drift is a different issue than data drift and is common to virtually all data science projects of sufficient complexity since features fed into the model may change. We have talked about schema drift before but post-deployment features can still change. It's interesting to note that there is no one size fits all solution to the problem of schema drift and actually there are various levels of schema drift. For example, you may have additive changes where you are only adding features and you may be able to simply set an option to allow the dataframe's schema to merge with the target table schema. You can implement this solution manually as well with a DDL SQL command like ALTER TABLE to avoid loss of historical data that may be required for model training.

However, what do you do when the order of columns changes, the data types are incompatible, for example, string to a float (casting a string to a float would lead to a loss of information), or there are other destructive operations on the table schema that are fundamentally not compatible with the target table? In this case, you may have to drop the table entirely. Many databases have a CREATE OR REPLACE TABLE statement, but you can implement this yourself by checking if the table exists and if it does dropping it and then recreating it. You should be careful to use atomic operations though if you're deploying this code to a distributed environment since race conditions and a source of strange errors in production are possible.

Traditional ways of handling schema drift like slowly changing dimensions don't work well for data science since features can change rapidly and the entire table structure may change this actually has consequences for model training since you could risk wiping out historical data required at a future point in time for training the model so running an ALTER TABLE statement on specific columns may be the safest bet along with regular backups of the data if possible. The schema itself, the metadata that describes the data types and structure of the table, needs to be stored as well with each version since of course this will change as well.

The Need for a More Flexible Table in Data Science

We talked about schema drift, and if you have actually worked as a data scientist, you might have encountered the problem of features being added or subtracted, names changing, data types changing regularly, and having to constantly update your tables. Traditional wisdom in database management assumes that the table structure is fixed which doesn't work well for data science.

While No-SQL databases and columnar storage address the problems of having a more flexible API and how to store data for analytical queries, you still need to handle schema drift. It's interesting to note that traditional SQL is based on relational algebra, and the equivalence with relational calculus under certain conditions such as domain independence is known as Codd's theorem.

While relational algebra and relational calculus are equivalent, relational calculus focuses on what to retrieve rather than how to retrieve it and so is more flexible. In relational calculus, there is no description of how to evaluate a query but instead a description (very similar to a prompt) of what information needs to be retrieved.

Whether a new kind of database is needed for data science that can better handle schema drift on a foundational level while maintaining performance for analytical queries is an open question that remains to be answered, but perhaps this technology could come from a fusion of large language models and the decades of wisdom built into relational database engineers ("optimizers"). In the next section, we will take all of the pieces of the lifecycle we have learned so far and discuss model deployment in general and ways to integrate all of the pieces of the puzzle into an existing business ecosystem.

Model Deployment

The model needs to be integrated into an existing business process. This seems like a technical problem but largely depends on your organization and industry. In the next section, we will look at how you can integrate your model into your business process as part of a larger system involving people, processes, and technology.

Deploying Model as Public or Private API

In the previous chapter, we talked about inference pipelines and microservices but for simple use cases where you only want to deploy a model, so it can be consumed as a private or public API endpoint, and there are many cloud services for doing this type of task; these types of cloud offerings are often called model as a service or inference as a service.

Hugging Face, for example, provides inference endpoints to easily deploy transformers, diffusers, or custom models to dedicated fully managed infrastructure in the cloud. This offering is a platform as a service where Hugging Face handles the security, load balancing, and other low level details, and you can choose your cloud provider and region if you have data compliance requirements. You can also choose public or private endpoints (intra-region secured AWS or Azure PrivateLink to VPC) that are not accessible over the Internet.

Integrating Your Model into a Business System

For stakeholders, one strategy for hedging against the vicissitudes of business is improving operational efficiency, reducing costs, and identifying opportunities to improve decision-making processes through models and innovate on insights found through data science. However, integrating a model into an established business system is a challenging problem and one that might be glossed over by data scientists and other technical leaders. The challenge is magnified when the machine learning system requires input from multiple departments and teams within those departments that have conflicting goals.

One way you can start to bring a model into an established business environment is by thinking incrementally and identifying opportunities where machine learning could bring the most value. Start with the pain points; are there repetitive tasks that could be automated? Are there tasks that nobody wants to do and may be a quick win? A good example would be cleaning data. Nobody enjoys data cleaning, but it's a business necessity and if that process is still done in spreadsheets, it may be a good candidate for automation.

Once you've identified a task that could be automated, you can investigate the data sources and look for ways the process could be improved. At this stage, it's critical to create a strategy and secure stakeholder buy-in for the first few phases of your project. Once you've proven you can bring value, adding multiple data sources, testing algorithms, training models, adding monitoring, and alerting can naturally add value and provide a segway into the next phases of the project.

Developing a Deployment Strategy

There are several established frameworks for data science and data mining that you might want to consider when building a business strategy and executing against that strategy. Although this book is meant to cover the technical aspects of the MLOps lifecycle, model deployment involves people and processes, and having a set of tools for execution can serve as a kind of checklist and mitigate risk of forgetting a step. Here are a couple frameworks you might incorporate into your own model deployment strategy.

CRISP-DM: The CRISP-DM (Cross-Industry Standard Process for Data Mining) is a standard framework originally developed for data mining but applies equally well to machine learning and data science. One of its advantages is it applies across multiple industries (we will look at specific case studies in a later chapter but it's worthwhile to have a framework that applies to multiple industries in mind). It has six phases which broadly correspond to phases in the MLOps lifecycle including model deployment:

1. Business understanding

2. Data understanding

3. Data preparation

4. Modeling

5. Evaluation

6. Deployment

Each of these phases consists of tasks, and phases follow sequentially with arrows between data preparation and modeling (unlike the spiral we talked about earlier), but this gives a very structured approach to data science and you can use it as a deployment checklist to make sure you're not missing steps. For nontechnical stakeholders, this may be a good way to communicate the various phases in a linear way. In the next section, we will look at ways in which these frameworks can be used to reduce technical debt.

Reducing Technical Debt in your Lifecycle

Technical debt can appear in many forms and can come about in different ways from working too fast to using suboptimal algorithms to forgetting how code works and making changes without updating old code and documentation. At the model deployment phase, it's critical to have standards in place to reduce technical debt across all stages of the lifecycle. Here are some deployment checklists you can use to ensure you're paying down technical debt in a timely manner:

1. Implement quality checks and linters before deployment. Friction between teams can often be reduced by doing something as simple as installing a code linter to ensure code is formatted in a standard way, eliminating arguments over what style is the best (since most data scientists have their own style). This can be done, for instance, on the main branch of the shared repository your team uses.

2. Hold regular code reviews and designate someone in charge of merging PRs (or you could implement this in a round-robin style).

3. Periodically reassess model performance post-deployment, and keep up to date with alerts and errors that are generated once the model goes to production.

4. Automate testing and monitoring as much as possible.

By following some of the preceding strategies, you can minimize technical debt post-deployment. At this point, you might be asking, I've already deployed my model, I've set up monitoring and automated testing, is it all hands-off from here? The answer is unfortunately, no. Data changes, environments change, and this does not stop after you deploy your model. Remember, the lifecycle is a continuous process. In the following section, we will look at what this process and how you can apply Agile principles in data science to make the process more efficient for you and your team. One way that you can reduce technical debt is with generative AI. In the next section, we will briefly look at how you can use generative AI to reduce technical debt by automating code reviews.

Generative AI for Code Reviews and Development

Generative AI leverages large language models which use reinforcement learning and the quadratic complexity of the transformer architecture at scale to billions of parameters. It can automate common tasks in coding and provide feedback through prompt-based development. With the rise of tools like AutoGPT, even prompt engineering is slowly being replaced.

Will this be a good thing for data scientists? I think so, as it can automate the boring stuff. Even for software developers, if you're a creative builder, you will be able to be more productive.

One way generative AI could improve the data scientist development cycle is through automated code reviews, getting feedback on their code before it is deployed. Some other ways but this list is by no means exhaustive.

- Automated code reviews

- Optimizing code (focus on accuracy)

- Translating between SQL and Python or other languages (removes translation bottlenecks)

- Generating tests for test-driven development

However, focus needs to be on validation. Output of generative models cannot be trusted, and data scientists will play a vital role in ensuring the validity, accuracy, and quality of model output when generative AI is mis-used. We should also be mindful of the cost per token and the license requirements before using this in your data science development cycle. We'll talk a lot more about these ethical issues in the next chapter, but generative AI has potential to reduce technical debt and free up time for doing data science.

Adapting Agile for Data Scientists

You may have heard of Agile before especially if you have worked on software projects in the last 20 years. Agile is a project management methodology, and although some developers have a love-hate relationship with Agile, some principles can be adapted to data science, and others fail miserably.

One principle of Agile that can be adapted is the principle of regular communication. Data science and software development share a common thread in that it's really all about communication. There's even a term for this called Conway's law which states

Any organization that designs a system will produce a design whose structure is a copy of the organization's communication structure

What does this mean? It means, if you fail to communicate effectively with other data scientists, the structure of your project will reflect this chaos and the system you end up creating will be a mess. Setting up a side channel for communication, for example, via Microsoft Teams, can help to dissipate this risk.

How about another conundrum often faced in real-world projects: conflicting requirements. In data science, stakeholders can be incredibly demanding, and requirements you gathered in the early phase of the lifecycle will change and you will face conflicts. We can borrow from another principle of Agile development to help in this scenario, namely, prioritizing tasks.

Another important principle of both agile and other methodologies like Twelve-Factor App is the notion of clean code that can be run in separate build, run and deploy stages and that can be easily adapted. While clean code and code management is an important tenet of Agile, in data science, it's all about the data. By emphasizing proper data management, we can ensure our models are accurate. For example, developing and prioritizing a process to improve the consistency of our labeled data set could have a massive impact on the accuracy of our models.

You may want to consider applying test-driven development to data science especially in early-stage development. While testing data heavy workflows is not easy, choosing a test framework, for example, Pytest or Hypothesis, and developing data fixtures (preferably ones that use realistic data from a database) can ensure models and code are performing as you expect even after you've deployed your model. These tests from the early development stage can easily be added to a CI/CD pipeline as well and become part of the model deployment process.

One area where Agile fails miserably in data science is regular sprints. Since the lifecycle has feedback into previous phases, for example, retraining models due to data drift may require reengineering some of the features or collecting more data. How do you anticipate these changes and fit them into a regular sprint? This is a difficult question as urgent and important tasks can get added to the board mid sprint and cause havoc on data science teams. Understanding the difference between model-centric and data-centric workflows may help to align teams and reduce some of the pain points in trying to pigeonhole data science into regular sprints.

Model-Centric vs. Data-Centric Workflows

When we talk about model deployment, there are two main approaches we could take to the overall process: a model-centric approach and a data-centric approach. What do I mean by model-centric and data-centric?

In order to illustrate this somewhat philosophical concept, let's suppose you are working on an NLP problem. You're trying to classify unstructured data collected from a free form response field on a survey into categories that can help the support team quickly prioritize issues. For example, the text "I have a problem with my Internet connection" may be classified as "connectivity issue". In fact you've done an exploratory analysis of the data, and you know 90% of the training data falls into the buckets of "connectivity issue," "hardware issue," and "authentication issue," corresponding to labels in your training set. However, there's a lot of training data, several gigabytes, and there's some ambiguity. You also had to label a lot of the data by hand, and you're not sure if it's completely consistent, and 10% of the data may be classified into new buckets.

Your model accuracy is only 70%, and you decide to improve this accuracy by changing the type of model and its architecture. Eventually you decide to try transfer learning, fine-tuning the last layers of a large language model on your specific data set, and this improves the accuracy to 85%. Since you primarily focused on the model and how you could improve the model, you've taken a model-centric approach to this problem, but is there another way?

In fact you could have taken a data-centric approach to improving model accuracy by focusing more on the data. You could have improved the consistency of the labeled data or developed a process to label the 10% of the data that was unlabeled. This would have been a data-centric approach.

In reality, you could mix the two and use semi-supervised learning, developing a model or rule to label the remaining 10% of data and then working to increase the consistency of the data, taking both a model- and data-centric approach to improving model accuracy.

So which approach is better? There is some debate, and both approaches can have their advantages and disadvantages, but when the problem is well-understood, optimizing the model makes sense. When the problem is less well-understood, there's ambiguity or complexity in the data or we're working with a very large amount of data, and then a data-centric approach may be the best option since the focus would be on capturing the variability and complexity of the data to move performance metrics in the right direction. Like most things in engineering, there are guiding principles and rules of thumb but with no clear-cut answer that applies universally in all cases. Regardless of the approach you take for your specific problem, you will want to automate the process of deployment as much as possible, and in the next section, we'll borrow from a DevOps concept of continuous delivery and continuous deployment and see how it applies to stochastic systems.

Continuous Delivery for Stochastic Systems

In the previous three chapters, we discussed several types of pipelines. We talked about ETL and ELT pipelines for refreshing our feature stores, we talked about training pipelines and actually built an end to end training pipeline, and we talked about inference pipelines that automated the model prediction. How about model deployment? Is there a pipeline we can create for this process? The answer is yes and it even has a name; these types of pipelines are called CI/CD pipelines.

CI/CD (continuous integration and continuous deployment) are a type of pipeline with several steps to guarantee that each time there's a code, model, or data change, that change gets tested and deployed to the right environment.

You may have several types of environments including development, testing, staging, and production consisting of databases, configuration, code, and data that need to be deployed to these environments. The CI/CD pipeline will consist of several steps:

1. *Version control:* The pipeline can be "triggered" whenever there's a commit to the main branch. This could, for example, be a pull request after a code review and cause the pipeline to start.

2. *Automated tests:* After starting, several tests will be run. As a data scientist, you can define what tests get run; for example, you may want to check if your features have the statistical properties you expect before deploying. These tests can include security tests, data quality and code quality checks, as well as formatting like linking.

3. *Build step:* After the tests have passed, the next step of the pipeline will take the code, data, and models and package them up into an environment and runtime. This may be a docker container, for example, which can be deployed.

4. *Deployment:* Once the changes are packaged and containerized, the container is deployed to the target environment. This environment could be production, releasing your changes to a live environment with end users (do you see why we need tests first?)

Introducing to Kubeflow for Data Scientists

Kubeflow is an open source tool for data scientists that makes it relatively straightforward for both data scientists and MLOps engineers to build, deploy, and manage workflows at scale. Kubeflow provides several features for deploying models that are particularly useful for data scientists like Jupyter notebook server (similar to the one we build ourselves) for managing and deploying models and code.

Kubeflow is designed to work on top of Kubernetes, so it may be overkill for your project. In the lab, you'll be able to optionally remove the Kubeflow step of the CI/CD pipeline if you only want to say host your project on GitHub or push your code and models to a docker container and host it on a docker registry. However, knowledge of Kubeflow is worthwhile for data scientists because you may encounter it in the wild, and knowing that a tool exists that abstracts away some of the details required to work with Kubernetes is enough to get you started on the right path.

For data scientists, you can use Kubeflow in several ways to do machine learning at scale.

1. Kubeflow provides a Jupyter notebook server for developing and test models. In combination with MLFlow, this can be a powerful tool for setting up experiments and hyper-parameter tuning

2. *Scaling workflows:* Data scientists can leverage Kubeflow to do model training at scale. Kubeflow can be used to provision resources like clusters required for distributed training on GPUs or CPUs and takes care of scheduling, orchestration, and managing cluster resources.

3. *Model deployment and serving:* Data scientist can use Kubeflow to deploy models to production and serve production models to end users by deploying them as Kubernetes services (remember, this is for a full-blown application or inference API). You can manage or fine-tune the Kubernetes deployment as well add load balancers and other services so you can scale up or scale down to match demand.

I've also said this several times before; but, for some projects using Kubernetes, it is not necessary, and you may choose a batch oriented workflow for the deployment step in which case you only need to build a batch inference pipeline and use your model to make predictions in batch. This is a completely valid way to deploy models. In the lab, we'll look at creating a CI/CD pipeline that you can modify to match your particular deployment needs.

Lab: Deploying Your Data Science Project

This is the final hands-on lab of the book, and you're going to build your own CI/CD pipeline. The goal of this lab is to have a CI/CD pipeline that is part of the toolkit and that you can modify to deploy your own projects to the cloud by adding steps as necessary. We'll be using GitHub actions in this lab. You can follow along with the following steps to understand how the pipeline is constructed or look at the finished CI/CD pipeline located in the .github folder of the final MLOps toolkit included with this chapter.

Before you proceed with the lab, you should know that YAML is another data format for configuration files consisting of key value pairs that can be arranged in a hierarchy. It's a human-readable format (actually it's a superset of JSON) and is a widely used standard for defining infrastructure as code, CI/CD pipelines, and a range of other configurations used in MLOps.

1. Create a new GitHub repository for your data science project (if you need help with this, refer to the lab from Chapter 3 on setting up source control).

2. Create a .github/workflows folder in the project root. In our case, this folder already exists.

3. Create a new YAML file in the .github/workflows folder and name it, for example, cicd_model_ deployment.yml.

4. Edit the YAML file as needed for your specific data science project. For example, you may need to update the name of the Docker image and the name of the container registry or remove the step to deploy to Kubernetes if you are not using Kubernetes with your project.

5. Commit the changes and push these changes to the repository created in the first step.

6. Add two secrets to the repository settings: REGISTRY_USERNAME and REGISTRY_ PASSWORD. These secrets should be kept confidential and correspond to the username and password for the container registry (e.g., Docker Hub or Azure Container Registry) that you are using.

7. Try to push changes to the main branch of the repository; the pipeline will automatically be triggered.

This CI/CD pipeline performs the following steps:

- Checkout the code from the repository.

- Set up a Python environment with the specified version of Python.

- Install pipenv and the project dependencies.

- Convert notebooks to Python scripts.

- Run Pytest to test the code.

- Build and push a Docker image with the latest changes.

- Deploy the Docker image to Kubeflow using kfctl.

You can modify this lab to fit your needs; you now have a full CI/CD pipeline with automated tests and a way to deploy your models to Kubeflow whenever a change is pushed to main. Remember, you should go through a PR process to ensure code quality before pushing to main. We've also included all of the notebooks from previous labs in the toolkit as a complete package.

Open Source vs. Closed Source in Data Science

Machine learning software can be open and closed, and if you've been following industry trends, there is a battle between the two philosophies as companies seek to establish a data moat; the open source community continues to develop open source versions of tools, models, and software packages.

Somewhere between the two are composed of both open and closed components (maybe we could refer to this as "clopen" software). This is further complicated by models that significantly transform the input like generative AI. When deploying models that use open source components you have to make a technical decision whether to open source or closed source your software at the end of the day and which components you choose and accompanying licensing impacts this decision. This adds even more complexity to the problem of MLOps placing a premium on MLOps practitioners to make ethical decisions when it comes to the decision systems they are deploying, regardless of the underlying technology behind the models. In the next chapter, we'll look at some of these ethical decisions and how they impact the MLOps role.

Monolithic vs. Distributed Architectures

Architecture is about trade-offs, and although we've covered many rules of thumb in this book like SOLID principles, distributed architectures for more event driven and real-time workstreams vs. batch oriented architectures that tend to be more monolithic, there is no one perfect architecture for each project, and you need to understand the trade-offs and the type of performance, security, data, and process requirements to decide what is the best architecture. Once you are committed to an architecture or platform, it can be difficult to change though, so you should do this ground work up front and commit to one type of architecture and platform.

Choosing a Deployment Model

In data science, there are several types of deployment models that can be used, and in some cases, you need to support multi-model deployment. Choosing a deployment model that best suits your needs is key to a smooth transition from development to deployment.

On-premises deployment: In this deployment mode, you utilize your own servers or IT environment (physical hardware, e.g., your own GPU enabled server running Jupyter labs). Although this gives you maximum control over the hardware, you are responsible for patching, updates, any upgrades, and regular maintenance as well as the inbound and outbound network connectivity and security.

Public cloud deployment: This may be a cloud service provider like Azure, for instance, with your own resource groups and cloud services such as Databricks. Cloud deployment may also include public services like releasing packages to PyPi or hosting packages on web servers or even GitHub.

Mobile deployment: Creating machine learning for smartphones or mobile devices is becoming more popular lately. Since these devices have limited memory compared to servers, you need to choose between hosting your models in the cloud and connecting to them from the device or reducing the size of the model. There is ongoing research to reduce the size of large language models and other models, for example, quantization[1] (representing the model weights as fewer bits) and knowledge distillation ("distilling a model[2]") to achieve a smaller size.

---

[1] Kohonen, T. (1998). Learning Vector Quantization. In *Springer series in information sciences* (pp. 245–261). Springer Nature. https://doi.org/10.1007/978-3-642-56927-2_6

[2] Yuan, L., Tay, F. E. H., Li, G., Wang, T., & Feng, J. (2020). Revisiting Knowledge Distillation via Label Smoothing Regularization. https://doi.org/10.1109/cvpr42600.2020.00396

Post-deployment

The post-deployment, although not technically a phase since it's an ongoing continuous process so it is not formerly part of the lifecycle, refers to the stage where the trained model is deployed to production and being used. Some of the considerations during this phase are communicating with stakeholders, soliciting user feedback, regular maintenance, and monitoring (e.g., of an API you rely on is deprecated and it must be updated, or if you find a CVE or common vulnerability exposure that impacts a PyPi package you're using, you need to patch it).

Beyond security and stakeholder feedback, collecting user feedback and monitoring how the users are interacting with your model can be invaluable for future projects and be used to train the model and augment existing data sources. Post-deployment monitoring ensures that all of the models deployed to production continue to provide business value and are used in an ethical way.

Deploying More General Stochastic Systems

Can we use the principles in this book to deploy more general stochastic systems such as Bayesian machine learning models? The answer is yes, but we should discuss some of the caveats.

If you use a library like PyMC3 (we used this in the second chapter to create a Bayesian logistic regression model), you can still save your model, but you should choose a custom serialization framework to match the model architecture, for example, ONNX, an open standard for neural network architectures, but others include HDF5 and Python's pickle (e.g., this works well with Bayesian models from PyMC3).

You may also need to consider the types of performance metrics you want to track, for example, Bayesian information criterion for feature selection or Bayesian credibility interval along with a prediction.

The other problem is you'll need to carefully consider sampling methods you use and may have to have hardware to ensure you have sufficient entropy for random sampling. Some of the algorithms may not scale well to large data sets or be intractable, so you may have a need to use Monte Carlo methods as opposed to a "big data" solution that may be a necessary approach for some algorithms.

There may be other stochastic algorithms that you may encounter that need to productionize. For example, reinforcement learning could be applied as part of a training pipeline to do hyper-parameter search or in specific use cases in healthcare, finance, and energy to simulate physical systems and make recommendations, dynamic planning, and natural language processing.

If you use a reinforcement learning algorithm like Q-learning, you will have to think about how to represent your environment and agents and how to update a Q-table and choose a framework that can handle interacting with the environment between learning steps, so you may choose a framework like Ray RLlib framework that offers support for highly distributed workflows.

Understanding the problem type may help you to identify the frameworks available since you should not reinvent the wheel (e.g., reinforcement learning frameworks, deep learning frameworks, frameworks for Bayesian inference). Other times, you may be able to achieve similar results with another approach where a library of framework exists (e.g., many problems can be reframed to use a different methodology, like how you can solve the multiarmed bandit problem using reinforcement learning or Bayesian sampling, and this is a kind of equifinality prosperity of many stochastic systems).

Still, you may one day encounter a bespoke stochastic algorithm that has never been used before in the wild, where no Python wrapper exists, and in that case, you would have to build your own from scratch. In this scenario, you would require knowledge of a low level language like C++, compilers, hardware, distributed systems, and APIs like MPI, OpenMP, or CUDA.

Summary

In this chapter, we looked at the spiral MLOps lifecycle and its different phases. We took another look at reducing technical debt from a holistic point of view after understanding each phase of the lifecycle. We discussed the philosophy behind taking a model-centric vs. a data-centric view of MLOps and why when working with big data, a data-centric view that encapsulates variability and complexity in the data may be preferable. We took a look at continuous delivery for stochastic systems and how we could adapt principles in this chapter to deploying Bayesian systems or more general types of stochastic systems along with some of the technical challenges. Finally, you did a hands-on lab, designing a CI/CD pipeline for the final toolkit that is a part of this book. Here is a summary of some of the topics we covered.

- Introducing the Spiral MLOps Lifecycle

- Reducing Technical Debt in Your Lifecycle

- The Various Levels of Schema Drift in Data Science

- Model Deployment

- Continuous Delivery for Stochastic Systems

In the final two chapters, we will diverge from the technical and hands-on components and instead take a deep dive into the ethical considerations around using AI and machine learning responsibly. We'll focus on model fairness, bias reduction, and policy that can minimize technical risk.

CHAPTER 8

Data Ethics

The **panopticon** is a design that originated with the English philosopher and social theorist Jeremy Bentham in the eighteenth century. The device would allow prisoners to be observed by a single security guard, without the prisoners knowing they were being watched. Today, the panopticon is used as a metaphor to highlight the threat to privacy and personal autonomy that comes with the collection, processing, and analysis of big data and shows the need to protect personal information in the face of increasing technological advancement. For example, multinational businesses face increasing scrutiny over how to store, process, and transfer private user data across geographic boundaries[1].

In this chapter, we will discuss data ethics (derived from the Greek word *ethos* meaning habit or custom); the principles that govern the way we use, consume, collect, share, and analyze data; and how as practitioners of data science can ensure the decision systems we build adhere to ethical standards.

Although this might seem like a diversion from the technical into the realm of applied philosophy, what separates the data scientist from traditional software engineers is that we work with data and that data can represent real people or it may be used to make decisions about entire groups of people, for example, loan applications or to deny someone a loan.

---

[1] Increasing uncertainty over how businesses transfer data across geographic boundaries is a current issue in data ethics.

© Dayne Sorvisto 2023
D. Sorvisto, *MLOps Lifecycle Toolkit*, https://doi.org/10.1007/978-1-4842-9642-4_8

If we don't consider what type of data goes into those decisions when we train a model, we may be responsible for building systems that are unethical. In the age of big data, where organizations collect vast amounts of data including demographic data which may be particularly sensitive, how that data is collected, stored, and processed is increasingly being regulated by policies and laws like the GDPR data protection act in the European Union and similar legislation in countries.

You may be a technical wizard at statistics or data analysis or software development, but without a solid understanding of data ethics, you may be doing more harm than good with your work, and your work may end up being a net negative to society as a whole. It is my opinion that what separates a scientist from a nonscientist or an engineer from a nonengineer isn't just knowledge but ethics. In this chapter, we will give a definition of data ethics and clarify some of the guiding principles you can use to shape your technical decision-making into ethical decision-making.

While this is not a book on generative AI, due to recent events, generative AI is set to shape a lot of the regulation around data ethics in the coming years. We will also cover some of the ethical implications of generative AI in this chapter, so you can understand the implications to your own organization if you incorporate generative AI into your data science projects.

Finally, we'll provide some recommendations for you how you can implement safeguards in your data science project such as retention policies to mitigate some of the risks that come with working with PII and other types of sensitive data.

Data Ethics

Data ethics is a branch of applied philosophy concerned with the principles that distinguish "good" decisions from "bad" decisions in the context of data and personal information. Unlike morality, which may determine individual behavior, ethics applies more broadly to a professional set of standards that is community driven.

Some of the ethical questions that data ethicists are concerned with include the following:

- Who owns data, the person it describes or the organizations that collect it?

- Does the organization or person processing the data have informed consent (important in the healthcare industry)?

- Are reasonable efforts made to safeguard personal privacy when the data is collected and stored?

- Should data that has a significant impact to society as a whole be open sourced?

- Can we measure algorithmic bias in the models we use to make decisions?

All of these questions are important as data scientists since we have access to vast amounts of personal data, and if this data is not handled properly, harm can be done to large groups of individuals. For example, an application that is meant to reduce bias in human decision-making but then exposes personal information of the groups it's trying to help may end up doing more harm than good.

Avoiding algorithmic bias and discrimination, improving the transparency and accountability of the data collection and analysis process, and upholding professional ethical standards are critical to the long term success of data scientists and MLOps and ensuring your models are valuable and sustainable.

Model Sustainability

I want to define the concept of model sustainability which I think is valuable to keep in mind when considering the role data ethics plays in data science and technical decision-making. What does it mean for a model to be sustainable? To be sustainable, it needs to adapt to change but not just technical change, change in society as a whole.

The fact is some data is political in nature; the boundaries between data and the individuals or group's data represents can be fuzzy, and when we start adding feedback loops into our model and complex chains of decision-making, how our models impact others may be difficult to measure. The other problem is social change is something not often considered by technical decision-makers, and a lot of software engineering is creating methodologies that protect against technical change but not social change. As data scientists, we need to be cognizant of both and have methodologies for making our models robust to social change as well which may come in the form of regulatory requirements or internal policy.

So how do we decide whether our model is able to adapt to regulatory requirements and social change? In the next section, we'll define data ethics as it applies to data science and discuss many issues around privacy, handling personal information, and how we can factor ethical decisions when making technical trade-offs.

Data Ethics for Data Science

How can we improve our ethical decision-making? In the real world, you may encounter trade-offs; for example, you may have data available that could improve the accuracy of your model. You may even have a functional requirement to achieve a certain accuracy threshold with your model. However, it's not acceptable to increase accuracy at the cost of algorithmic bias in the model. It is simplistic to assume just because a variable is "important" from a prediction point of view that it should be included automatically. This is also part of the reason why feature selection shouldn't be fully automated.

There are many ways to monitor bias, and this should be a part of the continuous monitoring process at minimum. A plan should be made to reduce algorithmic bias either by finding substitutes for variables that are sensitive, removing them all together. How the data was collected is also important; if the data was inferred without the user's informed consent, then it may not be ethical.

It's also possible stakeholders may not understand the implications of algorithmic bias in a model or the ethical implications of using sensitive PII in a model. In this case, it's the responsibility of the data scientist to explain the problem just as they would any other technical blocker.

Since data ethics is a rapidly evolving field, there are laws and regulations such as GDPR that can provide guidelines for making ethical decisions. In the next section, we will cover some of the most common legislation from around the world that may have impact to your projects.

GDPR and Data Governance

The General Data Protection Regulation (GDPR) is Europe's latest framework for data protection and was written in 2016 but became enforceable on May 25, 2018, replacing the previous 1995 data protection directive. The GDPR document has 11 chapters around general provisions, data rights, duties of controllers and processors of data, and liabilities for data breaches. One of the biggest impacts of the GDPR is its improvements in the way organizations handle personal data, reducing organization's ability to store and collect personal data in some circumstances and making the entire data collection process more expensive.

Personal data is any data which identifies or could identify a person and includes genetic data, biometric data, data processed for the purposes of identifying a human being, health-related data sets, and any kind of data that could be discriminatory or used for discriminatory purposes.

As a data scientist, if you do business with clients located in the European Union, you may have to abide by the GDPR. How does this translate into technical decision-making? You will likely have to set up separate infrastructure for the storage of data using a data center that is physically located in a particular geographic region. You will also have to ensure that when data is processed and analyzed, it does not cross this boundary, for example, moving data between geographic zones may have regulatory implications.

Similar legislation has been passed in several other countries since the GDPR such as Canada's Digital Charter Implementation Act on November 17, 2020. Although GDPR is a general data protection regulation (hence the name GDPR), there are regulations that apply to specific industries especially in healthcare and finance.

HIPAA: HIPAA or the Health Insurance Portability and Accountability Act is a 1996 act of the US Congress and protects patient data and health information from being disclosed. Since HIPAA is an American law, it only applies to American companies and when working with American customers, but if your organization does business with US citizens, you need to be aware of this law. The equivalent legislation in Canada is PIPEDA (Personal Information Protection and Electronic Documents Act) and is much broader than HIPAA, covering personal information in addition to health and patient data.

Ethics in Data Science

There are some guiding principles data scientists can use to make more ethical decisions. These principles include the following:

- Identify sensitive features and columns in a database and apply appropriate levels of encryption to PII (personally identifiable information).

- Set up mechanisms to decrypt PII if necessary but ensure that appropriate security and access controls are in place such as row level security and that only the information necessary to a job is made available.

- Add continuous monitoring to identify bias in model output for demographic data using metrics such as demographic parity.

- Assess the data set to understand if sensitive information could be inferred from any of the attributes, and take measures to remove these attributes or put in place appropriate safeguards to ensure this information is not misused.

- Understand how the models you develop will be used by business decision-makers and whether your model introduces any kind of unfairness or bias into the decision-making process either through the way the data is collected and processed or in the output of the model itself.

While these are not an exhaustive list, it should serve as a starting point for further discussion with your team to set standards for ethical decision-making and to highlight the importance data ethics plays in our own work. In the next section, we will look at an area that poses some risk for data scientists: the rise of generative AI.

Generative AI's Impact on Data Ethics

In 2023, Databricks released an ai_generate_text function in public preview that returns text generated by a large language model (LLM) given a prompt. The function is only available with Databricks SQL and Severless but can be used, for example, when creating a SQL query against a feature store. A data scientist could use this function to add generative AI to their project, and this is only one early example of how generative AI is increasingly making its way into data science tools.

The risk of being incorrect when discussing an event that is currently unfolding is relatively high, but this chapter wouldn't be complete without discussing the impact generative AI is having on data ethics. One of the biggest challenges generative AI poses to data ethics is related to data ownership.

How generative AI will impact how we view data ownership is still speculative as of 2023, but observers are already starting to see the profound impact it is having. A lot of the debate is around whether a human input into a model still owns the output of that model after it is sufficiently transformed. This is an incredibly interesting question that is poised to disrupt a lot of the current thinking that exists around data ownership, and if you use generative AI in your data science project, you need to be aware of the implications. I would suggest for the time being at least label output generated by a generative AI so you can identify it in your code base if you need to remove it in the future. Setting up a tagging system to this would be a clean way to implement a strategy in your own organization.

Safeguards for Mitigating Risk

We could spend years studying data ethics, and we still would never cover every scenario you might encounter. A compromise is needed between theory and practice to allow the reality of working with sensitive data attributes and PII and planning for the worst-case scenario such as a data breach or misuse of this information. Here are some safeguards you can implement in your own data science projects to mitigate this risk.

- Implement data retention policy, for example, removing data that is older than 30 days.

- Only collect data that is necessary to the model at hand and don't store data that is not relevant to the model especially if it contains PII.

- Encrypt all features that are considered PII such as email addresses, account numbers, customer numbers, phone numbers, and financial information such as credit card numbers.

- Consider implementing row level security and using data masking for tables and views that contain PII.

- Rotate access keys regularly and ensure data is encrypted in transit and at rest using latest encryption standards.

- Check PyPi packages and third party software before using them in a project in case they contain malicious software.

- Work with the security team to create a plan to monitor and protect data assets and minimize the risk of data breach.

- Implement continuous bias monitoring for models that use demographic data to ensure that the output is fair.

- Consider tagging anything created with generative AI during development.

Data Governance for Data Scientists

Data governance in the context of data science refers to a set of policies, procedures, and standards that govern the collection, management, analysis, processing, sharing, and access to data within an organization. Data governance is vital to provide guarantees that data is used responsibly and ethically and that decisions that come about as a result of a data analysis whether it be an ad hoc analysis or the output of an automated system are reliable, accurate, and ethical and are well-aligned with the ethical goals of the organization.

Data quality management is a part of data governance that implements data quality checks to ensure data is reliable. This goes beyond basic data cleaning and preprocessing and may include business initiatives in master data management and total data quality to maximize the quality of data across the entire organization rather than within a specific department.

Data security is another component of data governance and ensures that it is protected from unauthorized access and that the organization is taking steps to mitigate the risk of data breach. Policies such as requiring de-identification, anonymization, and encryption of data systems both at rest and in transit may be enforced by the data governance and security teams depending on the organization's threat model. The role of a data scientist and MLOps practitioner is to ensure the policies are implemented in accordance with these policies and to provide recommendations on how to mitigate risk of data breaches. Unfortunately, many data science tools are not secure, and malicious software is all too common in PyPi packages. A common attack is changing the name of a PyPi package to a name used internally by a data science team and hosting the malicious software on a public PyPi server. Such attacks are only the tip of the iceberg because security is often an afterthought in analytics and not a priority, even in Enterprise analytics software that should come with an assumption of security.

Data stewardship is another area of data governance related to data ownership but is more concerned with defining roles within different data teams such as data analyst, data engineer, MLOPs, and data scientist. In a RBAC or role based access control security model, each role would have well-defined permissions and responsibilities that can be enforced to protect data assets.

Finally, an organization should have a document defining and describing its data lifecycle. We talked about the MLOps lifecycle, but data also has a lifecycle, as it's created, and it's transformed into other data, creating new data sources, and these data sources are used but ultimately at some point are either deleted or archived and stored long term (requiring special consideration in terms of security). This entire lifecycle should be a part of the data governance process within your team to minimize risk of data loss and data breach and guarantee the ethical use of data across the entire data lifecycle.

Privacy and Data Science

Privacy concerns arise in the collection, storage, and sharing of personal information and data sources containing PII as well as in the use of data for purposes such as surveillance, voice and facial recognition technology, and other use cases where data is applied to identify individuals or features of individuals.

The history of data privacy can be traced back to the early days of computing. In 1973, some of the first laws on privacy were created with the passing of the Fair Credit Reporting Act which regulated the use of credit reports by credit reporting agencies to ensure not only accuracy but also the privacy of customer data. The following year, the United States also passed the Privacy Act which required federal agencies to protect privacy and personal information. Similar laws were passed in Europe in the 1980s, and by the 1990s with the rise of the Internet, data privacy concerns became an even bigger part of the public conscience.

Data science teams should only collect data that is necessary to the model at hand or future models and should ensure they have consent from the users whose data they're collecting. Not being transparent about the data collection process or how long data is stored means a risk to the reputation of the organization.

How to Identify PII in Big Data

When we're working with big data sets, these can be big in terms of volume but also in terms of the number of features, so-called "wide" data. It's not uncommon to have hundreds or even thousands of features.

One way to identify PII is to write some code that can dynamically churn through all of the features and verify columns like "gender," "age," "birthdate," "zip code," and any kind of demographic features that doesn't uniquely identify a person. While primary keys may be an obvious type of PII if they can be used to identify a person (e.g., a customer account key), for other features whether or not they can identify a person may require some more thought.

You may have to do some math around this; for example, if you have a combination of age or birth date and zip code, you might be able to identify a person depending on how many people live in a certain zip code. You could actually go through the calculation by using the Birthday Problem that states in a random group of 23 people, the probability of 2 people in that group having the same birthday is 0.5 or 50%.

We could generalize this heuristic and ask for any subset of demographic features in a big data set: What is the probability of a pair or combination of those features uniquely identifying a person? If the probability is high, you may have a hidden ethical trade-off between using the feature and increasing accuracy of your model and dropping the feature from your data set. At what specific threshold is acceptable to you and your problem depends on the problem, how the model is used, and your strategy for handling the PII.

This illustrates two important points when identifying PII: It's the combination of features that might uniquely identify someone rather than any one feature on its own, so when you're working with big data sets in particular, this is something to consider. Additionally, we can mathematically quantify the risk of identifying a particular person in a data set in some circumstances and actually quantify the risk.

Using Only the Data You Need

We've talked about PII but also there's a common sense approach here: We should only be using the data we actually need for the model at hand. Given, there may be an auxiliary need to use demographic data for marketing purposes, and that may be the reason why you need to include it in your feature set, but as much as possible, you should try to trim the fat and reduce the amount of data you're using. This also helps with performance; you don't want to bring in ten columns that are not needed since that's going to be a waste of space and bandwidth.

One question you can ask to trim the fat is are the features correlated with the response variable? This is a relatively common sense approach and may not work in all situations, but identifying the variables in your model that have no correlation with your target variable(s), you can create a shortlist of variables that could be removed. In the next section, we'll take a step back and look at data ethics from the point of view of data governance and the big picture impact of our models on the environment and society.

ESG and Social Responsibility for Data Science

Social responsibility and ESG (environmental, social, and governance) are increasingly becoming a part of organization strategies and future goals. Since data scientists seek to unlock value in data, understanding ESG and the role social responsibility plays in their organizations' long term goals will become increasingly important to the role of data science and MLOps.

Social responsibility in data science means the use of data to make decisions that benefit society, promote social good, and prevent harm to individuals or groups of individuals. An example is that data can be used to identify patterns of bias in big data and inform decision-makers on how these patterns of bias can be reduced. In industries such as energy, ESG involves a more concrete tracking of carbon emissions and the impact of the business on climate change and the environment and is a new opportunity for data scientists to drive positive change by coming up with innovative ways to measure ESG impact and make ESG initiatives data-driven.

Data Ethics Maturity Framework

If you remember way back in Chapter 1, we defined the MLOps maturity model and discussed different phases of maturity and how we could evaluate the maturity of a data science project. We can develop a similar framework for ethics in data science based on many data governance maturity frameworks used across industries.[2] Take a look at Table 8-1.

---

[2] Al-Ruithe, M., & Benkhelifa, E. (2017). *Cloud data governance maturity model.* https://doi.org/10.1145/3018896.3036394

Table 8-1. *Data ethics maturity framework*

Dimension	Questions	Definition	Level 1	Level 2	Level 3
Privacy	Is personal data collected, stored, and used?	Personal Identifiable Information (PII) is defined as information related to an identifiable person(s) such as email address, credit card, and personal address	Physical locks or physically protected PII, for example, in a safe or locked cabinet. PII may exist on paper	Digitization and digital safeguard such as RBAC or ACL. File level encryption or database encryption applied to PII columns	Both technological and policy safeguards. Administrative policy and governance to protect privacy
Bias (nonstatistical)	Do models make different decisions for different demographic groups? Could these decisions translate into unfair treatment for different groups or is the model fair?	Bias refers to different model output for different demographic groups and not bias in a statistical sense	Bias identified in model but no concrete safeguards	Bias identified and has been analyzed and measured. Plan to reduce bias in models	Continuous monitoring and active bias reduction in models

(continued)

Table 8-1. (*continued*)

Dimension	Questions	Definition	Level 1	Level 2	Level 3
Transparency	Is the data science team transparent about how data was collected?	Data transparency refers to data being used fairly, lawfully, and for valid purposes	What data sources are used is clearly documented	Business is aware of all data sources used and who has access to it	Individuals and businesses know which data sources are being used and data integrity is protected

How might you use Table 8-1 on your own project? Although we could add more dimensions such as social responsibility, data accessibility, and data security, understanding the impact of your data and models on transparency, privacy, and bias is a good starting point for understanding the ethical considerations around your problem.

Why a framework? It may seem like overkill but can help you to reduce technical risks associated with unethical use of data by providing a measurable and pragmatic method for evaluating and monitoring the project across these various dimensions.

This framework is not theoretical, and the process should also start early before data is collected since collecting and storing PII may already violate laws and regulations without having to have processed it. While transparency and privacy may also be qualitative dimensions that can't be measured directly, bias (not bias in the statistical sense) meaning whether the model is fair or not actually can be measured quantitatively using metrics like demographic parity (a kind of conditional probability). How you measure bias is different for each type of problem. For example, for a multi-class classification problem, you might compute bias differently than for regression, but continuous monitoring and active bias reduction would be what differentiates a level 3 and level 2 solution in this maturity framework. In the next section, we'll look at responsible use of AI and how some of the ideas around AI might apply to data science as a whole.

Responsible Use of AI in Data Science

Data science is not artificial intelligence, but data scientists may use AI such as generative AI both as developers to make themselves more productive and to generate features for even entire data sets. For example, one application of generative AI is you can sample from a generative model to "query" it, and this might be as simple as feeding in a prompt but could actually involve complex statistical sampling methods with applications from recommendation to data augmentation.

While the applications of generative AI in data science are without bound, there are ethical challenges posed by generative AI, and these are multifaceted challenges at the intersection of society, technology, and philosophy. It is not even known at this time whether emergent properties such as consciousness itself could arise from certain types of AI, and this creates a moral quandary.

With increasing attention on the responsible use of AI, the ethics of artificial intelligence is becoming a mainstay in many data science discussions across all types of industries and organizations, even those not traditionally seen as technology companies.

Topics like bias in large language models whether or not language models or other types of AI can have emergent properties like consciousness and the existential threat posed by AI will continue to push our understanding of data ethics.

Staying on top of the rapid advancements in AI almost requires a superhuman AI itself to digest the vast amounts of information available, but there are some resources available.

Further Resources

Data ethics is a rapidly evolving field and multidisciplinary field at the intersection of technology, society, and philosophy, so it's important to stay current. Some ways you can stay up to date with data ethics include the following:

1. Subscribing to industry publications including journals, magazines, and blogs and following news, for example, setting an alert for GDPR or similar data regulation. This may help to stay up to date on current debates and emergent news.

2. Joining reputable professional organizations. Although there is no centralized body for data ethics, finding like-minded professionals can provide guidance and practical experience that you may not find elsewhere especially if there is controversy around a particular ethical question.

3. Taking courses and reading the history of data ethics can help to make more informed decisions when working with data and personal information.

Data scientists and MLOps professionals that understand data ethics will help to standardize this body of knowledge and keep the data ecosystem free from long term negative consequences of making unethical choices when working with data.

In the final lab for this book, we will look at how you can integrate practical bias reduction into your project to reduce the risk of unethical use of the models you create, and the lab will provide some starter code you can use in your own project.

Data Ethics Lab: Adding Bias Reduction to Titanic Disaster Dataset

If you've done any kind of machine learning, you're probably familiar with this Titanic data set, but it's always struck me how people go through the example without thinking about the types of features used in the example, so it's always felt incomplete. In the lab, you'll add the necessary code to compute demographic parity to decide if the model is fair or not using the Shap library.

Here is the recipe for this lab.

Step 1. You'll first need to install the Shap library from PyPi preferably in your virtual environment.

Step 2. Run the python file chapter_8_data_ethics_lab.py.

Step 3. Decide if the model is biased or not based on the demographic parity. Feel free to change the data to make the model unbiased.

Summary

In this chapter, we defined data ethics and discussed why ethics are important for professionals that work with data including the following:

- Ethics for Data Science Projects

- GDPR and Data Governance

- Ethics in Data Science

- Further Resources

We looked at some guiding principles for applying ethical decision-making in data science and some case studies and examples of specific regulation that governs the ethical standards within the data ecosystem today. While technology and ethics are extremely important, it is only half the picture, and both regulation regarding data ethics and the technology we in MLOps are spared by domain knowledge and the intricacies of each individual industry. In the next chapter, we will look at specific industries from energy to finance and healthcare.

CHAPTER 9

Case Studies by Industry

In this chapter, we will finally look at the most important aspect of data science: domain knowledge. You can think of this chapter as providing a kind of ladder, as you won't be a domain expert reading this chapter alone but maybe you will be able to get to where you want to go by asking the right questions. After all, data science is not about technology or code, but it's about the data and, more specifically, the domain knowledge and concepts that data represents. Each industry has unique problems that may not be well understood outside of that industry.

So what is data? Data in some sense is more general than even numbers since it can be both quantitative and qualitative. Numerical data is a type of data, while categorical or natural language-based data represent raw concepts. Domain and industry knowledge is really the "soul" behind the data, the elusive part of any data science project that gives the data meaning.

While many modeling problems are considered solved especially in supervised machine learning where basic classification and regression problems can be repurposed over and over again possibly with very little domain knowledge, to actually drive performance metrics and solve novel problems that will bring a competitive advantage to your industry and even to be able to identify which problems are actually "hard" will require domain knowledge.

© Dayne Sorvisto 2023
D. Sorvisto, *MLOps Lifecycle Toolkit*, https://doi.org/10.1007/978-1-4842-9642-4_9

This is the last chapter in the book because unlike previous chapters, domain knowledge is not easy to learn; it has to be earned through years of experience. Mathematics can be learned and technology can be learned to some degree, but domain knowledge is purely experience based, knowing what to measure, how to measure it, what is noise, what features to throw away and what to keep, how to sample the data to avoid bias, how to treat missing values, how to compute features (code that encodes all of this business logic to eliminate bias in data can get very complex), and knowing what algorithms are currently used and why for the particular domain are all something that needs to be learned from experience. It also changes, and unless you're working in a domain, it can be hard to even get an understanding of what problems are important and what models are considered solved.

It's the goal of this chapter to discuss some of the problems across different industries and look at ways in which MLOps can improve the lives of domain experts in those areas or to provide some further information for data scientists that have experience in one industry and are looking to transfer that knowledge to another industry. We'll also take a look at how we can use this knowledge, store and share it across industries, and leverage it for strategic advantage for our organization by using the MLOps lifecycle.

Causal Data Science and Confounding Variables

One of the things that makes data science difficult is confounding variables and illustrates why we need to have domain knowledge to truly understand the dependence structure in our data. Without a solid understanding of our data, we won't be able to identify confounding variables, and we may introduce spurious correlations into our results, and our models won't be an accurate representation of reality.

What is a confounding variable? A confounding variable is a third variable that influences both the independent and dependent variables in a model. This is a causal concept meaning we can't just use correlations to identify confounders, but we need a solid understanding of the data domain and the causal factors that underlie the model; after all, correlation does not imply causation. A visual representation of a confounding variable is shown in Figure 9-1.

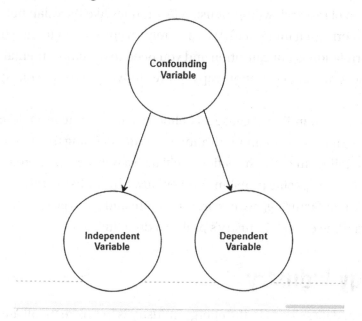

Figure 9-1. *A confounding variable influences both the independent and dependent variables*

For example, you might look at the impact of a variable like amount of alcohol consumed on a daily basis on mortality rate. However, there are many confounding variables like age that may be considered confounding variables since age can have an effect on both alcohol consumption and mortality rate. Another good example is the placebo effect where you might have a randomized experiment with two groups but one group was administered a placebo. Since the belief that the treatment is effective may influence the outcome of the treatment (e.g., believing a treatment will give you more energy may cause you to feel less tired), the experiment needs to be controlled to account for the placebo effect by splitting participants into two randomized groups.

So how can we identify confounding variables? There is no known algorithm for learning all of the cause and effect relationships and identifying confounding variables (although causal data science is an active area of research with promising techniques like Bayesian networks and counterfactual inference). We can't rely on correlation to help us here since correlation is not causation and you need to develop a mental model of the cause and effect relationships you're studying to truly avoid spurious results.

This understanding of cause and effect can only come from domain experience and is the main motivation for understanding the domain we're modeling. In the next section, we'll take a bottom-up approach and look at domain specific problems broken down by industry from energy, finance, manufacturing, healthcare, and more and try to get a better understanding of each industry's problem domains.

Energy Industry

One use of data science in the energy industry is in upstream oil and gas. Geostatistics, which takes into account spatial dependencies in data (data points that are close together on the Earth's surface are assumed to be similar), leads to important applications like Kriging and geospatial sampling.

Collecting data on well reservoirs is costly, and having techniques that can infer unknown data without directly measuring it is an important application. Another area is in midstream oil and gas, where energy needs to be physically transpired in pipelines or, in the case of utilities, physically transported over a distribution network. How do we detect anomalies and make this process more efficient?

Safety is also critical in this area, and using data science to identify leakages and other anomalies to reduce outages or, in the case of oil and gas, to prevent shutdowns is a focus of a lot of modeling applications.

Manufacturing

Manufacturing is becoming increasingly data-driven as organizations recognize the potential of data analytics to optimize their operations and increase operational efficiency. An increase in operational efficiency of just 0.1% can translate into large cost savings in absolute terms since that 0.1% is relative to EBITDA or operating income.

The use of data science and MLOps can help manufacturing executives gain insight into production processes, reduce waste, develop lean processes, improve quality control, and forecast equipment and component failures before they happen.

A concrete example is predictive maintenance or forecasting the time to failure or similar variable from sensor data. Sensor data may be from entire fleets of equipment or production equipment from manufacturers and can help predict which components are likely to fail. This may aid in scheduling maintenance, reducing downtime, and increasing output.

Statistical quality control is another area of operational research where data science can lead to innovative improvements. Identifying trends in product defects can help manufacturers analyze root causes of defects and adjust their processes to reduce future defects or to adjust inventory levels and lead times on the fly.

Transportation

Transportation and manufacturing have many overlaps in terms of data science use cases. Transportation itself is a huge industry that is best broken down into different subindustries that include railways, shipping, aviation, and more. Striking the right balance between safety and efficiency is one of the drivers of using data science in transportation, and again predictive maintenance has many applications.

Edge devices that may be attached to entire fleets of vehicles emit various sensor readings like pressure and temperature (we looked at an example of this in our feature engineering lab) and can be analyzed to forecast time to failure and improve scheduling efficiency. It's important to note here that these sensor data sets are massive data sets especially if they come from entire fleets of vehicles and may include real-time data, so MLOps play a critical role in transportation data science especially in scheduling off-peak hours and minimizing scheduling conflicts and complex route planning.

For example, how do you determine the best routes to take when your loss function includes information on fuel consumption and other transportation costs and you have to minimize this loss over a massive data set of sensor readings? To make matters more complicated, the fleet of vehicles may extend across broad geographical boundaries and include different units that need to be normalized, and all of this data has to be processed in a way that takes into account operational safety as well as efficiency. Safety data itself can be multimodal coming from traffic cameras, sensors, and other sources to identify areas with high incidence rate.

Retail

Retail is an industry that has been completely transformed by data science in the last 20 years, from recommender systems and customer segmentation to pricing optimization and demand forecasting for new productions.

Customer segmentation (looking at the customer along dimensions like geography, psychology, demographic, buying patterns, purchasing preferences, and other traits) helps to personalize messages and product offerings and can be used in combination with a recommender system.

Demand forecasting can help retailers analyze historical sales and transactional data to study variables such as weather, promotions, holidays, and macroeconomic data to predict demand for new productions and decide where to allocate resources and marketing efforts. Demand forecasting is primarily used for reducing inventory levels and increasing sales by anticipating spikes in sales volumes. Demand forecasting can also be applied to optimize supply chains by adapting to spikes and decreases in demand.

Related to sales, price optimization is a classical use case for retail data science. Data science can help optimize pricing strategies based on current market trends like inflation and interest rates. Customer demand forecasts can also be fed into these models with competitor pricing to maximize profit margins year over year. Developing pricing strategy is necessary in competitive markets like retail where pricing may be a key differentiator of the product.

Recommender systems can be created by querying models to recommend new products and services to customer segments based on past purchases (when available) or other data like browsing history for online retailers.

Agritech

Agricultural communities developed over 10,000 years ago, so it's not every day agriculture gets a major overhaul, but data science has tremendous potential in the intersection of agriculture and technology called agritech to improve agricultural processes and increase yields and overall efficiency in precision farming.

Precision farming in particular uses data science to collect data from sensors, drones, and other sources to optimize crop yields and reduce use of fertilizers and pesticides that can harm the environment and reduce yield. The main factors that influence crop yield include weather, agricultural land, water, and harvest frequency, and data collected from sensors can be used to maximize yield.

Finance Industry

Finance is one of the most interesting industries for data science applications. Fraud detection (a kind of anomaly detection problem) seeks to detect fraudulent transactions and prevent financial crime. Fraud detection is possible in part because of our ability to process vast amounts of data and to measure a baseline behavior in the transactions to detect patterns of fraud even if they're less than 1% of the entire sample.

Risk management is another area where data scientists build predictive models, to quantify risk and the likelihood and frequency of occurrence. Predicting which customers might default on a loan, for instance, is an important problem in predictive modeling. MLOps can help to streamline risk management problems by bringing transparency and explainability into the modeling process, introducing mathematical methods like SHAP or LIME to report on which attributes went into a particular loan decision. Model explainability and fairness are particularly important in credit risk scoring where demographic features (income, geographic location), payment history, and other personal information are fed into the model in hope of getting a more accurate picture of someone's credit risk at the current time.

Metrics like net promoter score and customer lifetime value are frequently used in modeling. Industry standards are extremely important in the finance industry especially when it comes to data science. Risk modeling, for instance, is important because if we can calculate risk of churn or risk of default, we can concentrate resources around preventing those customers from churning provided they have sufficient customer lifetime value.

However, how you approach the risk model in finance may be different than other industries. Continuous features are often discretized, meaning the features are placed into buckets. One of the reasons for this is so we can create a scorecard at the end since there are often laws and regulations around reporting credit risk and the models need to be interpretable by someone without advanced knowledge of the model. There's also the assumption of monotonicity with credit risk. This is difficult to include in some models.

Healthcare Industry

We can look at applications of data science in healthcare and predict where MLOps can impact healthcare. One area that is very active is in medical image analysis and, more generally, preventative medicine.

Preventative medicine uses X-rays, CT scans, MRI scans, and healthcare data to detect abnormalities and diagnose disease and malities faster than a human doctor or even a traditional lab test could. Imagine you could diagnose disease years in advance and treat them before they become a problem that threatens the health of the patient.

While X-rays, CT scans, and other types of medical imaging would require computer vision models such as convolutional neural networks, preventative medicine may also look at the entire history of the patient to summarize it for medical professionals (e.g., autoencoders or topic analysis algorithms) require natural language processing and domain knowledge of medicine. These will be vast data sets and require infrastructure to support big data as well as require security and data privacy safeguards to protect patient data (e.g., this data may be regulated by HIPAA or similar regulation).

Predictive analytics can ultimately be used to reduce hospital readmissions and reduce patient risk factors over the long term, but moving these metrics requires reporting, monitoring, and ability to feed patient outcomes back into the model for retraining.

Two emerging areas of research in healthcare are drug discovery and clinical decision support systems. Causal inference can be applied to discover new combinations of drugs and build new treatments or even speed up the clinical trials or augment data sources that are too expensive to collect.

Monitoring infrastructure can be set up to monitor patients in real time and provide healthcare practitioners with real-time data on patients that can be used to make better healthcare decisions resulting in better patient outcomes and reduced hospital visits. The potential to increase the efficiency and optimize resource allocation in the healthcare space will be one of the most important applications of MLOps in the twenty-first century.

Insurance Industry

If you're a data scientist in this field, then there's a lot of opportunity for innovation. One unique example is preventative maintenance. We might not think of preventative maintenance having applications in something like insurance, but what if insurance companies could use sensor data to predict when vehicles need to be maintained, preventing breakdowns before they happen and keeping claims at a minimum. This would benefit both the drive of the vehicle and the insurance company.

Data science is playing an increasingly vital role within the insurance industry, enabling insurers to make more accurate assessments of risk and personalize policies. Most people know there are large volumes of customer data available to predict the likelihood of claims behind made, for example, insurers could use customer demographic data, credit scores, and historical claim data and develop personalized risk strategy models.

Fraudulent claims are expensive to insurers and lead to bottlenecks and inefficiencies in the process as insurers seek to eliminate fake claims with strict policy rules and procedures for underwriting. However, analyzing patterns in customer data, we can use anomaly detection to identify fraudulent claims without the additional cost.

Customer experience is another area within the insurance industry that could use some improvements. Although we don't usually think of insurance companies as being in the customer service industry, with increasing competition in this space, using data science to fine-tune policy to customer needs would lead to new business opportunities. All of these types of models require big data, and MLOps can make the insurance industry much more operationally efficient, to automate the underwriting process and make personalized policy recommendations based on customer risk profiles and other factors.

Product Data Science

Each of the industries mentioned continue to be disrupted by innovative technology companies that are increasingly becoming data and analytics companies that leverage data to improve traditional business processes. A great example is in healthcare where machine learning is being applied to preventative medicine to diagnose disease and infections before they become advanced or untreatable. By developing new diagnostic techniques with machine learning, countless lives can be saved.

Another area ripe for disruption is in the financial industry where customers that would not traditionally qualify for a loan may be considered because there's data available to evaluate the risk of default.

While product data science is different in the sense that you need to understand the product end to end rather than building a model to make an existing process more efficient, your model needs to have product-market fit. Understanding the customer or end user of the group across various dimensions such as demographic, psychological, behavioral, and geographic data sets can help to segment customers and provide insight into what kind of model may best meet the needs of each customer segment.

Customer segmentation may be an invaluable approach to product data scientists and, also, the ability to ask questions to establish and uncover novel ways of modeling a problem since, unlike in industry, the modeling problem itself may not be a solved problem. This is why advanced knowledge of statistics and experience with research are required to be an effective product data scientist.

Research and Development Data Science: The Last Frontier

Data science at the edge is a rugged landscape, a mixture of many different disciplines that are constantly evolving. In fact, some industries may not even be invented yet. You might wonder how data science might look in 50 years. While predicting something like how data science will evolve 50 years out is clearly not possible, if we want to predict how we might better position ourselves to understand the massive amount of change in this field, we might want to look at research and development and the kind of impact data science has had on scientific research and business innovation.

In particular, we can look at areas from applied research to commercialization to new lines of business in various industries. Data science has increasingly become important in science and engineering, and although we can't predict the future, we can look at fields like genomics, neuroscience, environmental science, physics, mathematics, and biomedical research to gain an understanding of some global trends. We summarize these trends in the following.

- *Genomics:* Data science is used to analyze genomic and proteomic data to identify patterns, sequences, and mutations in genes and proteins. Deep learning systems like AlphaFold can accurately predict 3D models of protein structures and are accelerating research in this area.

- *Neuroscience:* Data science is increasingly being combined with brain imaging such as fMRI and EEG to unlock structure and function in the brain and provide new treatments for brain diseases.

- *Environmental science:* Data science is being used to analyze climate data, satellite imagery, and oceanographic and seismic data to understand how human activities impact our environment and to create new climate adaption technologies.

- *Physics:* Data science is used in physics to analyze big data sets and identify complex patterns in data from particle accelerators, telescopes, and scans of the universe. This information can be used to find new star systems and planets (data-driven astronomy) and to even develop new theories and models of the universe.

- *Mathematics:* Most of the focus on large language models has been on training these models to understand natural language and not formal languages like mathematics. While AI may not replace mathematicians completely, generative AI may be used to generate proofs, while formal verification may be used to validate these proofs. Building a system that uses both generative AI and formal verification systems like automated theorem provers as components will lead to groundbreaking results and the first proofs completely generated by AI mathematicians.

- *Biomedical research:* Data science is used to analyze clinical trial data, biomedical data, and data from drug trials to develop new treatments and interventions for debilitating diseases. Causal data science is an emerging area within data science that has tremendous potential to expedite biomedical research.

Although we can list many active areas of research where data science has had an impact or will have an impact in the future, this relationship goes in both directions. While branches of mathematics like statistics and linear algebra have had the biggest impact on data science so far, other areas of mathematics like topology continue to find its way into mainstream data science through manifold learning techniques like t-SNE belonging to the emerging field of topological data analysis.

Other areas of mathematics are slowly making their way into data science, and people find new ways to apply old mathematical techniques to data processing. One interesting area is in algebraic data analysis where age-old techniques like Fourier transforms and wavelets are being used to change the way we analyze data. I mentioned this in Chapter 2, but if you take the Fourier transform of a probability distribution, you get the characteristic function of that distribution. Characteristic functions are a kind of algebraic object, and they've been applied in many proofs in mathematical statistics like proof of the central limit theorem. While other applications of Fourier transforms like wavelet signal processing are being used in some areas of data science, there are many mathematical techniques that will eventually find their way into mainstream data science.

In the next few sections, we'll pivot back to more concrete use cases of data in industry and how you can apply them in your own organization.

Building a Data Moat for Your Organization

A data moat is a competitive advantage where data itself is treated as a business asset. By leveraging data as an asset, businesses can create barriers to entry for competition and use data as a strategic advantage. The key to building a data moat is using data in a way that cannot easily be replicated. As a data scientist, you know what data is valuable, but as an MLOps practitioner, you can use this knowledge to build a data moat.

The first step would be to collect as much data as possible but to implement quality gates to safeguard the quality of the collection. This may require an investment in IT systems and tools to collect and process data effectively to determine what should be kept and what is noise.

The next phase is to identify what data cannot be replicated. This is the most valuable asset for a business and might be customer data, industry data, or data that was extremely difficult to collect.

Once you have identified enough quality data sources that cannot be easily replicated, you can analyze this data to leverage it in your operations. The full MLOps lifecycle applies at this stage, and you may start with a single data science project and slowly, iteratively build toward becoming a data-driven organization where you can offer new innovative service lines and products from this data.

Finally, after you've integrated a certain level of MLOps maturity meaning you're able to create feedback into your data collecting process to create more data, insights, services, and products, you need to safeguard the data and protect it like any other highly valuable asset by implementing proper data governance policies. This entire process may happen over a number of years.

One of the difficulties in building your organization's data moat is lack of domain expertise and capturing domain expertise in your MLOps process. In the next section, we'll look at the history of domain experts and how organizations have attempted to capture domain expertise when building their data moats.

The Changing Role of the Domain Expert Through History

Throughout history, there have been many AI winters and many attempts to capture domain expertise and store it. Expert systems were formally around as early as the 1960s and were designed to mimic expert decision-making ability in a specific domain. The first commercial expert system called Dendral was invented to help organic chemists identify unknown organic molecules by analyzing mass spectra. This and subsequent expert systems were rule based, and by the 1980s, they were able to make use of some simple machine learning algorithms. In the 1990s, expert systems were used in industries ranging from finance to healthcare and manufacturing to provide specialist support for complex tasks, but there was a problem: These expert systems were difficult to maintain, requiring human experts to update the knowledge base and rules.

Today, chatbots use a different approach: generative AI creating new data from old data that are far less brittle than expert systems. However, currently there is no way to update these chatbots in real time (requiring layers of reinforcement learning), and if data used to train these models is insufficient, the knowledge will be inaccurate. These models are also expensive to train with a fixed cost per token, and you have to train one model per domain; there is little to no transfer between domains leading to knowledge silos. This leads to an interesting question: What is the role of the domain expert in data science in the face of this change?

A challenging problem is there are many different kinds of data scientists not just differentiated by role and skill but domain expertise: knowing what tools are useful and what problems have been solved before and being able to communicate that knowledge.

Mathematics is a universal language. Data visualization can be used to communicate results of data analysis but hides the details of how you arrived at that problem. To make things worse, each industry has its own vocabulary, standards, and ways of measuring. That being said, there are still a couple ways to store domain knowledge and maybe share that knowledge across industry and teams.

- *Documentation:* This is a straightforward way to store domain knowledge and share it across teams and industries. This may include books, technical manuals, research blogs from leading companies, academic journals, and trade journals.

- *Knowledge graphs:* Knowledge graphs are a way to organize domain knowledge into relationships between concepts. For domain knowledge that is highly relationship driven like social networks, this may be a good tool to represent knowledge.

- *Expert systems:* Expert systems were an attempt to represent expertise in a rule based system but have many limitations.

- *Ontologies:* Ontologies are a formal way to store academic domain knowledge by representing knowledge as a set of concepts and relationships between concepts. This differs from knowledge graphs in that ontologies are full semantic models for an entire domain while knowledge graphs are specific to a task.

- *Generative AI:* Large language models are increasingly being used to store domain knowledge. At this time, training large language models on custom data is an expensive process, but as the cost per token decreases over time, generative AI may become the standard way to share domain knowledge.

- *Code:* An example of this is the toolkit we created, but open source projects are good way to share knowledge across domains, for example, developing an R library to solve a problem in one industry and sharing it on CRAN so it can be applied in another industry.

- *Metadata:* Defining standard vocabulary for your industry and developing a metadata dictionary, for example, to annotate features in a feature store.

While we have many ways to share domain expertise from simple documentation to more formal methods to represent entire domains, sharing knowledge is only one piece of the puzzle. Data and the knowledge it represents grow over time and need to be processed not just stored. The situation where data outpaces processing capabilities may become a limiting factor in some domains.

Will Data Outpace Processing Capabilities?

IoT data is increasing at an alarming rate. This scenario is often called data deluge and happens when data grows faster than our ability to process it? While exascale computing promises to provide hardware capable of 10^{18} IEEE 754 Double Precision (64-bit) operations (multiplications and/or additions) per second, data is much easier to produce than it is to process. So-called dark data is produced when organizations have collected data but do not have the throughput to process it. While MLOps can help in unlocking some of this dark data, new systems, technologies, and hardware may have to be incorporated into the MLOps lifecycle to handle increasing volumes of data.

The MLOps Lifecycle Toolkit

The reader of this book is encouraged to use the MLOps lifecycle toolkit that is provided with the code for this chapter. I have added MLFlow and Jupyter lab components that use containers to the Infrastructure folder and added the model fairness code to the fairness folder so you can use it in your own projects. The accompanying software ("the toolkit") suit their own needs. The idea for an agnostic toolkit that can be used as a starter project or accelerator for MLOps can facilitate data science in your organization in combination with this book that serves as documentation for the toolkit.

Building a toolkit that is agnostic that includes tools for containerization, model deployment, feature engineering, and model development in a cookie cutter template means it's highly customizable to the needs of your particular industry and project as a way to share knowledge and provide a foundation for domain experts doing data science.

The field of data science is very fast-paced encompassing not only machine learning but nonparametric algorithms and statistical techniques, big data, and most importantly domain knowledge that shapes the field as a whole. As domain knowledge changes, the toolkit may evolve, but the invariants like mathematical knowledge, algorithmic thinking, and principles for engineering large-scale systems will only be transformed and applied to new problems. Figure 9-2 shows the end-to-end MLOps lifecycle components used in the toolkit as an architectural diagram.

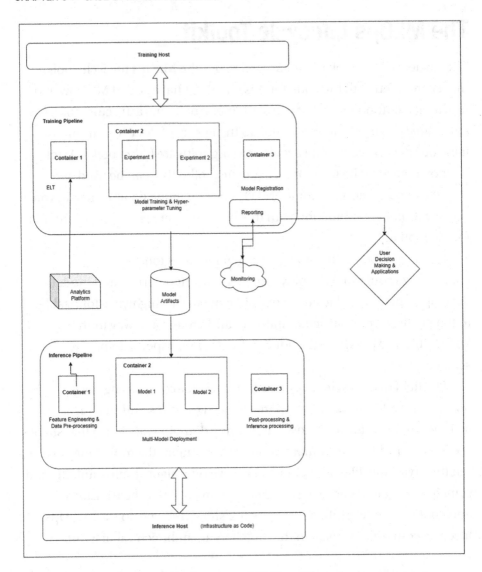

Figure 9-2. *Relationships between MLOps lifecycle components*

Summary

In this chapter, we looked at various applications of data science by industry. Some of the industries we discussed include the following:

- Energy Industry

- Finance Industry

- Healthcare Industry

- Insurance Industry

We concentrated on broad problems within each industry and emphasized the importance of industry standard techniques, vocabulary, and domain knowledge in data science. We looked at the changing role of the domain expert throughout history including attempts to capture knowledge and store it as expert systems and in generative AI. We discussed the hypothetical point where our ability to produce data may outpace our ability to process data and how this may impact the MLOps lifecycle in the future. Finally, we talked about contributing to the MLOps toolkit, the accompanying piece of software that comes with this book, providing the final version in the code that comes with this chapter with all the previous labs and infrastructure components used in previous chapters.

Index

A

Ad hoc statistical analysis, 9–10
Amazon Web Services (AWS),
94, 117–120, 142–143,
170, 199
Application programming
interface (API)
designing interfaces, 174
high-velocity data, 23
inference pipelines, 181–183
JupyterLab server, 93
keras, 64
model deployment, 199
multi-model deployments/
pulling models, 168
POST request, 179
PySpark, 63
RESTful, 178, 179

B

Bayesian model
approaches, 41
Bayes' rule, 40
frequentist model, 39
frequentist position, 41
frequentist statistics, 40

likelihood function, 40
logistic regression, 39
parameters, 40
probability, 39
trace plot, 39
Bayesian networks, 7, 38, 240
Building training pipelines
automated reporting, 150
batch processing/feature stores
feedback loops, 154
gradient descent, 150
LIME/SHAP, 153
mini-batch, 151
model explainability,
153, 154
online learning method,
152, 153
personalization, 153
stochastic gradient
descent, 151
ELT, 140–142
feature selection, 148
handling missing values,
145, 146
hyper-parameters
boosting model, 155
docker-compose file, 159